Pests of Ornamental Plants

THE AUTHOR

Dr. Sathe Tukaram Vithalrao [M.Sc., Ph.D., Sangit Visharad, IBT (Seri.), F.S.E.Sc., F.I.S.E.C., F.S.L.Sc., F.I.C.C.B.] is presently working as selected Professor in Entomology, Department of Zoology, Shivaji University, Kolhapur. He has teaching experience of 26 years in Entomology and 15 years in Agrochemicals and Pest Management at P.G. Level. He has written 30 books and published 215 research papers in national and international journals of repute. He has completed 4 major research projects and guided 17 Ph.D. students. He visited Canada (1988), Japan (1988), Thailand (2002, 2004), Spain (2005), France (2005), South Korea (2002, 2006) and Nepal (2007) for academic work. He is member of editorial board of eleven prestigious journals. He delivered 30 talks through All India Radio and internal conferences and involved in T.V. programmes on biodiversity of moths and butterflies. He published more than 35 popular articles in daily newspapers on insects and sericulture. He got several prestigious awards "Environmentalists of the Year-2003", "Bharat Jyoti", "Jewel of India", "International Gold Star", "Eminent Citizen of India", "Education Acumen", "Best Educationist", "Eminent Scientist of the Year-2008", "Lifetime Education Achievement" and "Lifetime Achievement in Entomology and Insect Taxonomy-2009". He has been also awarded several fellowships from different scientific and academic societies.

Pests of Ornamental Plants

T.V. Sathe

Entomology Division
Department of Zoology
Shivaji University
Kolhapur – 416 004

2017

Daya Publishing House®

A Division of

Astral International Pvt. Ltd.

New Delhi - 110 002

First Published, 2011
Reprinted, 2017

ISBN: 978-81-7035-757-5 (Hardbound)

Published by : **Daya Publishing House®**
A Division of
Astral International Pvt. Ltd.
– ISO 9001:2015 Certified Company –
4736/23, Ansari Road, Darya Ganj
New Delhi-110 002
Ph. 011-43549197, 23278134
E-mail: info@astralint.com
Website: www.astralint.com

Printed at : **Replika Press Pvt. Ltd**

CONTENTS

Chapter 1
INTRODUCTION

Gardening in India dates back to 2500 B.C. before Aryas came. Pipel *Ficus religiosa* was the first ornamental cultivation came from Mohen-jo-daro of the third millennium B.C. and Willow *Salix babylonica* from Harappa. The lotus has been frequently reported from vedic times. Atharva veda, Rigveda, Ramayana and Mahabharata compiled and mentioned garden trees/plants and flowers. The lotus was a popular flower and was regarded by hindus as symbol of purity. The great Emperor Asoka 264-227 B.C. has developed arboriculture as a part of state policy. During the rule of Chandragupta -II, Kalidasa mentioned several flowering plants like Parijata (*Nyctanthes arbor-tristis*), Madhavi, Kadamba etc. Ajantas (100-600 AD) are indicatives of gardens and flowers of ancient India. Before Mughals came to India there was a long blank in gardening history. However, Firoz Tughlaq (1351-1388 A.D.) developed about 1200 gardens around Delhi. The emperor

Babar was lover of flowers and gardens. He developed gardens at Panipat (1526 A.D.) and Agra. Oleader, white Jasmine and Hibiscus are some good names of flowers of the Babar regimes. Akbar (1556-1605 AD) developed Nasim Bag close to Dal lake and Tomb garden at Sikandra. Jehangir (1605–1627 AD) developed gardens at Shalimar, Achhabal and Veriang in Kashmir. Taj Mahal in Agra in the memory of his beloved queen, Mumtaz Mahal were the creations of Shah Jahan (1627-1628 AD). The best garden developed by this emperor is Shalimar garden in Lahore (Pakistan). Shershah Suri (1540–1544 AD) planted avenue trees on both sides of the famous Grand Trunk Road which he constructed during his short period of rule. Sikandar Bag was established by last Nawab Wajid Ali Shah (1847–1856 AD).

Several gardens have been established by the Rajput Rajas in India. Man Singh (1590–1615 AD) started the historic garden at Amber Fort was completed by Jai Singh II in 1699. The beautiful "Mandor garden" was established by Raja Abhai Singh near Jodhpur during 1724–1749 AD. Jai Singh II founded Jaipur city with a palace in the heart of city with outstanding gardens. The garden palace of Deeg was founded by Surajmal is one of the most beautiful gardens developed by Rajput King. Likely, gardens in Chittor are also outstanding creations in India.

Indian gardens were in symmetrical patterns but British changed into informal patterns. However, British brought many beautiful flowers in Indian gardens which refer to carnation, verbena, dahlia, larkspur, phlox etc. British also established some botanical gardens in India during their rule. There are several gardens in India.

Flowers and gardens have very important place in the life of nation and people. However, the potential of floriculture as an industry has not properly exploited in India. Floriculture can open up great opportunities to our poor farmers. Almost all kinds of flowers are grown in India at one or another areas as there are much diverse climatic conditions. China aster, Marigold, Carnation, Dahlia, Gerbera, Gladiolus, Jasmine, Chrysanthamum, Orchids, Rose, Tuberose, Cacti etc. have great commercial value. They should be cultivated on large scale because floriculture has tremendous potential in domestic marketing and at international scenario too. In India, 3,000 hectares of land is under floriculture. However, Horticulture in India is not flourished. One of the reasons for not taking floriculture on large scale in India is pests and diseases of ornamental plants. The problem of pests can be solved by studying pest distribution, marks of identification, pest biology, nature of damage, host plants and various control strategies including ecofriendly control. On this context the present book will add great relevance.

Gardens of India

There are uncountable gardens in India. But, some historic and very famous gardens are listed below with their specialities. Basically gardens are of two types:

1. Botanical gardens (for scientific studies) and
2. Pleasure gardens (for common citizens)

Botanical Gardens

1. *Government Botanic Garden, Ootacamund.* It is found in Nilgiri hills and designed with 6 sections and contains several species of plants.

2. *Botanical Garden, Coimbatore (TN)*: It is situated in Tamil Nadu Agricultural University. This young garden is established in 1908 and contains large number of plant species. Children's corner, green houses, rockery, Hibiscus varietal collection, large lawns and rosary are additional features of this garden.
3. *The Indian Botanic Garden, Sibpur, Kolkata (W.B.)*: It was established in 1787 which covers 190 hectares of land. It contains giant 200 year old banian tree and palms. Palm houses, orchid houses, ferneries house, several thousand herbaceous plants. The garden is having 15,000 trees and shrubs belonging to 2,500 species.
4. *The Agri-Horticultural Societies Garden, Kolkata*: This garden was started in 1825 and completed in good condition in 1872. Metalled paths in the garden, collection of Bougainvilleas, water lilies and large pond houses, rockeries, rich collection of creepers, climbers, a huge open lawn, children corner, auditorium, laboratories are important features of this garden. It is largest collection of trees, shrubs, creepers and some hybrids.
5. *Llyod Botanic Garden, Darjeeling (West Bengal)*: It is established in 1878 at the height of 2100 m from sea level at midst of Himalayas which comprises 24 hectares of land with beautiful terraces and metalled roads. The garden has 12 sections for various purposes. The garden represents about 1800 botanical species collected from Central Asia, Myanmar, Malaysia, Japan, America, Europe and Africa. It also contains rare and exotic conifers.

The garden distributes plants and seeds. 'Living Fossil Tree' and 'Ginkgo bilobo' are the attractions of the garden.

6. *National Botanical Research Institute, Lucknow*: It is famous as "Sikandar Bagh" which is extended into 27 hectares of land. It is established by Nawab Alikhan and improved by Nawab Waji Ali Shah (1789–1814). 'Sikandar Bagh' provides a fine lawns, rose gardens, conservatory, cacty house, lily pools, etc. The garden has its own library and departments of Botany, Chemistry, Tissue culture, Breeding Technology etc. for boosting research and training the man power in floriculture.
7. *Horticultural Research Institute, Saharanpur (UP)*: This garden was established before 1750. But, East India Company taken over this garden in 1817. It is famous for the collection of fruit trees, ornamental plants, shrubs, cacti and succulents. The garden conducts research on subtropical fruits.
8. *Rashtrapati Bhavan Garden, New Delhi*: Rashtrapati Bhavan is one of the largest building in the world established in 1929. Garden area is 15 acre bounded by paved red stone path. Two canals each of 5.40 m width run from North to South and two similar canals intersect these to form a 60 m square island in the centre. The garden contains quality and quantity seasonal flowers and has Mughal garden pattern. There are good collection of Bougainvilleas, bulbous plants and flowers, roser, orchids, cacti succulents, ferns etc. The garden remains open for about a month for general visitors.

9. *Botanical Garden, Forest Research Institute, Dehradun (U.P.)*: The garden was established in 1934 at an altitude of 663 m at Dehradun. It is extended in about 10 hectares of land and F.R.I. has 500 hectares land for Research and Training. The garden is objected for introduction of new plants from the world. Forest trees, ornamental plants, and economic plants is wealth of this garden. The garden has rich collection of indigenous flora, specially terrestrial and epiphytic orchids, ferns, shade-loving plants, cacti, succulents and also nurseries for providing seedlings to interested people or plant seed exchange programme. They have also an arboretum of 50 hectares containing forest plants.

Pleasure Gardens

There are several pleasure gardens in India which are developed by maintaining Mughal pattern. Important amongst them are given below:

1. *Mughal Gardens of Kashmir*: Shalimar garden, Nishat Bagh, Chasma-e-shahi, Verinag, Achabal, Bijbehara etc. are important gardens found in Kashmir. Planting of the majestic chenar trees in groups is most spectacular feature of these gardens.
2. *The Mughal Garden, Pinjore (Haryana)*: This garden was established by Fidai Khan in 17th century. It is on the Ambala-Shimla Road, Kalka. It is most beautiful and best preserved garden of North India. It is scattered over 25 hectares of land and divided into 6 terraces. There is central water channel, the water falls from one terrace to the other and into

the tanks. The tanks and water channels have numerous fountains. On other side of the central channel the paths are paved. Lawns and flowers beds, trimmed hedges, rows of bottle palms and many other ornamental shrubs and trees is attraction of this garden. Shish Mahal, Rang Mahal and Jal Mahal are the other attractions of the garden.

3. *Chandigarh Rose Garden*: Rose is king of the flower due to its beauty, elegance, grace and fragrance. Chandigarh Rose garden is one of the biggest rose garden in the world. It is established in 1966 and designed to contain about 60,000 roses and 5,000 outstanding cultivars. The garden is situated in the centre of Chandigarh city comprising about 15 hectares land. The garden has been laid in the most natural and informal manner as it is situated in a valley with natural streams. The stream adjoining area is with plantation of number of scented cultivars hence its entire area becomes full of fragrance. A section of moon light garden planted with pure white and scented flowers is one of the attraction to visitors.

   At present garden represents 1500 roses in the museum. Biggest rose garden of Newark is near New York in USA, it contains about 36,000 roses. The rose gardens in Oeste, Madrid, Spain contain 30,000 roses. The rose garden at Parque Bagatelle in Paris, France is also quite famous.

4. *Mandor Garden, Jodhpur (Rajasthan)*: It is one of the most beautiful garden of India which is situated 5 km away from the heart of Jodhpur, in desert

region of Rajasthan. It is very large garden in size and terraces have been constructed on the slopes of the hillocks. There are fountain fitted with coloured lights. The tea stall at highest terrace attracts the attention of every body. The garden is built by Raja Abhai Singh (1729-49A.D.).

5. *Sayaji Park, Baroda (Gujarat)*: Maharaja Sayajirao III established this garden in 1879. It is scattered in about 40 hectares of land and contains 8000 various ornamental trees. Arbour like bandstand, paved walks, lush green lawns, four fountains, illuminating changing multicolor lights and piped music are the outstanding features of this garden. Toy train, a giant wheel, children's park and traffic training centre for children are also attractions of the garden. Baroda museum, attached to the garden in 1894. A planetarium is recently added to the garden.

Rather than above mentioned, there are many more beautiful and outstanding gardens in India worth while visiting. In recent days, gardening has tremendous scope.

Chapter 2
PESTS OF ROSE

The rose belongs the family Rosaceae. There are about 120 species of roses in the world which are distributed to temperate and tropical zones of New Mexico, North America, North-east Africa, Himalayan belts of India and mountains of China and Japan.

For gardening following eight species have given priority:

1. *Rosa damascena*
2. *Rosa foetida*
3. *Rosa chinensis*
4. *Rosa gallica*
5. *Rosa gigantea*
6. *Rosa moschata*
7. *Rosa multiflora* and
8. *Rosa wichuraiana*

Wild growing species of Roses in India refer to:

1. *Rosa gigantea*
2. *Rosa odorata* Sweet var. *gagantea*
3. *Rosa leschenaultiana*
4. *Rosa foetida*
5. *Rosa clinophylla*
6. *Rosa involucrata*
7. *Rosa longicuspis*
8. *Rosa moschata*
9. *Rosa macrophylla*
10. *Rosa webbiana* and
11. *Rosa sericea*

The era of modern roses began with the crosses made between Chinese and European roses. The first hybrid of rose, "La Fance" came in existence in 1867 in France. The first yellow pernet rose was developed again in France. The first polyantha rose is cross between "Dwarf pink china" and *Rosa multiflora*" which produces numerous flowers at a time. "Foribundas" is recent hybrid which is very popularly grown with great interest was developed in 1912. The miniature roses are 'Pixie', "Baby Masquerade" and "Baby Gold Star". Cabbage roses, moss roses, musk roses, Australian Briars, Fritz Nobis, Sparrieshoop etc. are also commercially very important varieties at global scenario.

In Indian scenario Aruna, Belle of Punjab, Golconda, Scented Bowl, white nun, Delhi Prince, Temple flame, Delhi pink pearl, and many more have been developed for commercial use. Some scented roses include–Blue Moon, Sugandha (Red), Spartan, La France (Pink), etc. Scented miniatures are Pixie (white), Twinkles (white) and Midget (Carmine Red), while, Dorothy perkins (Rose pink), Honey moon (Golden yellow), Sympathie (dark red), Golden showers (Yellow), Albertine, coral Dawn etc are from climbers and Ramblers categories.

Some exhibition types of roses popularly cultivated in India are Avon (red), Golden Giant (Yellow), Maria Callas (Rose pink), Pusa Sonia (golden yellow), Rose Gaujard (bicoloured), Super Star (light vermillion), Virgo (white) etc.

Roses are cultivated with one color in single bed or with mixed colours. The best time for planting the roses is between September and October. In temperate zones it is done between October and April. Roses can be propogated by cuttings, Layering, Budding and grafting. In India prunning of roses is done during October.

Rose is King of the flowers. It is widely cultivated through out the world. In India, almost every state is involved in floriculture of rose. Rose flowers are in different colours *viz.*, red, pink, yellow, white and of light shades of different colours. Rose has tremendous export potential. Therefore, rose as floricultural international trade has great importance. Pure Indian varieties and hybrid varieties are two basic groups of roses. Pure varieties have typical pleasant smell is the attraction of everybody. Hybrid varieties represents no pleasant smell but they have different colors and size. Hence, hybrid varieties have also their own importance in floriculture business. Scent perfume industry is based largely on roses. Flowers are also used a food, Gulkand (a sweet). Both pure and hybrid varieties of rose get attacked by several pest insects and non insect pests. Important insect pests of rose are represented in Table 1 with their taxonomic position.

Table 1: Pests of Rose.

Sl.No.	Common Name	Scientific Name	Family	Order
1.	The Rose Aphid	*Macrosiphum rosaeformis* Das	Aphididae	Hemiptera
2.	The cotton aphid	*Aphis gossypii* Glover	Aphididae	Hemiptera
3.	The groundnut aphid	*Aphis craccivora* Koch	Aphididae	Hemiptera
4.	The white grub	*Holotrichia consanguinea*	Scarabaeidae	Coleoptera
5.	The white grub	*Schizonycha* sp.	Scarabaeidae	Coleoptera
6.	The caster hairy caterpillar	*Euproctis lunata* Walker	Lymantridae	Lepidoptera
7.	The grapevine thrip	*Rhipiphorothrips cruentatus* Hood	Thripidae	Thysanoptera
		Dendrothrips ornatus	Thripidae	Thysanoptera
8.	The caster slug	*Latoia lepida* Gram.	Cochlidiidae	Lepidoptera
9.	The citrus red scale	*Aonidiella aurantii* (Maskell)	Diaspididae	Hemiptera
10.	Rose leaf hopper	*Edwardsiana rosae*	Jassidae	Hemiptera
11.	The cotton grey weevil	*Myllocerus undecimpustulatus* Faust	Curculionidae	Coleoptera
12.	The Rose aphid	*Lindigaspis rossi*	Aphididae	Hemiptera

The Rose Aphid *Macrosiphum rosaeformis* Das

Distribution

Nilgiri Hills, Delhi, Punjab, Karnataka, Andhra Pradesh, Maharashtra, North India, South India.

Occurrence

In North India this species appears in the middle of November. Winged forms present from November to April.

Marks of Identification

December onwards population increases with peak in March. The Rose aphid *M. rosaeformis* is soft bodied, non winged form, measures about 2.5–2.6 mm in body length. They are with large red eyes and with black siphunculi. Abdominal tip of the insect is yellowish green, is characteristic of this species. However, the aphid forms found in South India show purple abdomen, dark legs and yellow head. Winged forms are also present.

Host Plants

Rose *Rosa* spp.

Life Cycle

The pest reproduced parthenogenetically and viviparously. The nymphal period is 11-14 days in non winged forms while in winged forms the nymphal period is 14-19 days, in March fastest development takes place. In north India the pest appear in November mid, the population peaked in March, later, declined in early April due to warm season. The winged forms multiplied fast from December onwards and peaked in March during which about 90 per cent are winged forms. However, most rapid multiplication takes place in late spring but can not withstand the summer due to host climate, resulting in decline in population.

Macrosiphum rosae (L.)

Distribution

World wide in distribution. This species is reported from Japan. However, it is not found in eastern Asia.

Marks of Identification

Apterous forms are medium sized to rather large, broadly spindle shaped, shiny, mid to dark green or deep pink to red brown or magenta coloured. They are with shiny black head and siphunculi. Antennae and legs are bi-coloured, yellow and black. Cauda is pale yellow. Black sclerites present on dorsal of abdomen. They measures about 1.7 to 3.6 mm. However, winged forms measures about 2.2 mm to 3.4 mm and they have distinct black sclerites along sides of abdomen.

Life Cycle

This is heteroecious holocyclic between Rosa and Dipsacaceae or valerianaceae in temperate regions. Some populations can be observed on rose during summer. In autumn migration to rose occurs on large scale. Later, this non migrant population produce sexuales (sexuparae) and many over-winter in mild-winter climate.

Vector Transmission

Transmit 12 plant viruses. PSV, Rose mosaic, Rose streak etc.

Macrosiphum euphorbiae (Thomas) (Figure 1)

Distribution

North America, New world wide. Recently spread in Central Asia and middle East.

Figure 1: *Macrosiphum euphorbiae.*

Marks of Identification

Non winged forms are shindle shaped or pear shaped with some shade of green or yellowish, pink or magenta. Eyes are distinctly reddish, Legs, cauda and bars have some coloration as body however, bars are darker towards apices, they measure about 1.7 to 3.6 mm in body length. Winged forms measure about 1.7 to 3.4 mm in body length and are with pale greenish to yellow-brown thoracic lobes.

Hosts

Rosa spp as primary host, polyphagus, 200 hosts are known from 20 families.

Virus Transmission

This species acts as vector for 40 non persistent viruses, BVN, DEM, PLR, Potato–LR etc.

Life Cycle

Reproduce parthenogenetically and sexually. This is heteroecious holocyclic in north eastern USA with wild or cultivated *Rosa* spp. as primary host plants. Sexual morphs are produced in small number.

Nature of Damage

Nature of damage of this species to *Rosa* plant is almost same as above species. Hence, it is discussed collectively. Both nymphs and adults suck the cell sap from tender leaves, buds and twigs of rose plant. While sucking the all sap, the pest inject toxins into the plant body resulting disfiguring and withering of flowers. Each aphid suck from the several places of the plant body making several punctures and wounds. Due to sucking cell sap the leaves become curly, petals become curty, and yellow, dry and drop down. The pest also secrete honey dew like substance on leaves and other parts. That invites saprophytic fungus to grow on the leaves and other parts, a thick black coating is generated which is called as sooty mould. The sooty mould affect photosynthesis of the plant by that the growth of the plant and the quality of flowers. The pest also acts as vector for transmitting several plant viral diseases as mentioned above. Aphids cause following problems to ornamental plant, Rose, in following ways:

1. Contamination: The presence of aphids in flowers and on leaves will reduce the value of flowers.
2. Feeding injury as discussed above.
3. Rapid development of pest and reinfestation.
4. Development of pest at cooler temperature.
5. Acts as vector for transmitting viral diseases to vegetables and ornamental plants.

Control Measures

Preventive

1. Collection and destruction of infected plant parts along with aphid populations.

2. Be alert for infection to plants while collecting healthy flowers.
3. Small white skins left on plants by aphids during the moulting process should be removed by hair brush.

Curative Control

1. *Mechanical Control*: Plants must be inspected frequently to find localized aphid infestions. After incidence, arrange yellow sticky traps for capturing aphid population. Winged forms will attract on large scale and will remain attached to sticky traps.
2. Green house covered with plastic that blocked UV light at wavelengths below 380 nm had fewer aphids than in green house covered with plastic that blocked UV light below 360 nm.

Chemical Control

Olympic products that are effective against aphids include:

1. Malathon 1 per cent G 60 wp and Malathon II
2. Azatin II
3. Triact 70
4. Olympic insecticidal soap

Mathon 1 per cent G and 6 wp are used to provide 8-17 weeks of aphid control on crops with longer production cycles.

Spray–	Phosphamidon	0.02 per cent (or)
	Malathion	0.03 per cent
	Rogor	0.03 per cent
	Azadirachtin	0.03 per cent

Biological Control

1. Lady bird beetles like *Menochilus* sp., *Coccinella* sp., feed on aphids.
2. Lace wing *Crysoperla* sp. feed on aphids.
3. Syrphids feed on aphids.
4. Some parasitoids cause mortality in aphids.

The Groundnut White Grub (*Holotrichia consanguinea* Blanchard)

Distribution

Widely distributed in India. This pest has been reported from Gujarat in 1957 on groundnut. Later, it has been reported from Haryana, Punjab, Rajasthan and Himachal Pradesh. Now it is widely distributed in several States of India including Maharashtra, Karnataka, Andhra Pradesh, Madhya Pradesh, U.P. etc. However, the pest has been first reported from Bihar State from Dalmia Nagar in 1909.

Marks of Identification

The grub is whitish in body colour with brown head. Full grown grub measures about 35 mm in body length. The grub is provided with 3 pairs of thoracic legs and is typically comma shaped. Adult is beetle with dull brown colour which measures about 18 mm in body length and 7 mm in width. The beetle causes damage only at night to leaves.

Host Plants

Rosa spp both Deshi and hybrid varieties are damaged by the beetle. The pest also feed on Groundnut, Sugarcane, Maize, Ber, Mango, Fig, Jowar, Brinjal, Chilli, Okra, Tomato,

Bornian, Neem, Ain, Pearl millet & grinea grass etc. It is polyphagus pest.

Life Cycle

The pest becomes active after the onset of monsoon. Adult beetles lay their eggs in the soil singly at the depth of about 10 cm. Eggs are rounded and whitish. They hatch within 7-10 days depending upon climatic conditions. The newly hatched grubs feed on roots of rose and other food plants with humus material. Newly hatched grub is about 12 mm long while, matured grub is about 53 mm long and 7 mm wide. The larval period is 8-10 weeks. Full grown grub pupates in the soil. The pupa is exarate type and measures about 21 mm in body length and is semicircular and creamy white. Pupal period is about fortnight. The newly formed beetle remain in the soil at the depth of about 10-20 cm at day time and come out at night for feeding purpose. Only one generation is completed during the year. The beetles formed in November remain in the soil till the next June.

Nature of Damage

Grubs feed on roots of the rose plant while, beetles (adults) feed on leaves of rose and other plants. The beetles come out from soil at night and feed on the leaves of rose by making numerous irregular, circular and semicircular holes (Figure 2), by keeping edge sides intact. The grubs feed on nodules of the fine rootlets ultimately killing the plant.

Control Measures

Preventive

1. Collection of eggs, grubs, pupae and adults and destruction of them.

Figure 2: Damage by Beetles.

2. Digging the field for exposing eggs, grubs, pupae and adults to natural mortality factors (Abiotic & biotic)
3. Beetles get attracted to light, they can be collected and destroyed.

Mechanical Control

Fluorescent light traps are very attractive to the beetles. During the warm and humid nights beetle emergence is high. They can be trapped and killed.

Cultural Control

1. Rose planting should be made in autumn for avoiding attack of white grub beetles.
2. Do not take rose continuously for several years. Sunflower is recommended as rotatory crop against above beetle.
3. At night beetle congregate on the trees such as neem, peepal, gular, mango, guava, Ain etc. shaking the trees and collecting the beetles

dropped on ground and killing them is good behavioural control.

4. Flooding field frequently can remove pest from the target area.

Chemical Control

1. Spray the crop with 0.1 per cent to 0.2 per cent Carbaryl or Monocrotophos 0.05 per cent at the time of mass emergence in June/July.
2. Dusting the crop with HCH 10 per cent 10 kg/ha.
3. Quinolphos S-G 2 to 2.5 kg/ha.

Microbiol Control

1. Use of spore dust of *Bacillus popilliae* mixed at 4 x 10 spores/kg and 2 x 10 spores/kg of soil around the root zone of rose crop.
2. Application of *Bacillus thuringiensis* viable spores as dust at 10 g containing 3 x 10^3 per gram.

Citrus Red Scales

1. *Aonidiella aurantii* (Maskell)
2. *Landigaspis rossi* (Maskell)

A. aurantii

Distribution

World wide in distribution. In India, it is very bad pest of citrus. Except Africa it has been reported from all citrus cultivating countries of the world.

Marks of Identification

The scales are brownish reddish in colouration, rounded/semicircular bodied, measuring about 2 mm in diameter. The adult female is flat, circular in shape, 2 mm in diameter with small antennae, rostrum which is twice the length of body. Males are elongated and winged. Thus, there is sexual diamorphism in this pest. Nymphs are flat bodied and smaller than adults. First instar nymphs are crawlers.

Host Plants

Rose, Citrus, Grape vine, Fig, Willow, Shisham (*Dalbergia sisoo*), *Acacia, Eucalyptus,* Ber, Banana, Guava, Jamun, Peach, Pear, Mulberry and many ornamental plants. Deshi varieties of rose are badly affected by scale insects.

Life Cycle

No egg laying outside the body of female takes place. Thus, instead of laying eggs, the female gives birth to nymphs. There is no parthenogenesis in this pest species. Female reproduces ovo-viviparously. However, some times, eggs are also laid. The pest is active throughout the year but, it is most active during the months August to October. The 1st stage nymphs are crawlers with well developed legs for moving and settling on the suitable plant part for sucking the cell sap. The pest likes tender leaves and tender stems for settlement and for sucking the cell sap. The female nymph can moult two times. Each nymphal stage lasts for 10-20 days. Finally, the nymph molt into adult.

Females are wingless and males are winged individuals. The pest reaches to its sexual maturity within 10-15 weeks. Females can survive for several months by feeding on cell sap of rose but, males (adults) can not survive for longer period. They fertilize the females and die. The pest can complete several generations in a single year.

Nature of Damage

Both nymphs and adult females suck the cell sap from leaves, flowers and stems of the rose plant. However, Branyovitis (1953) says that these insects spend more time in moulting than in feeding. They suck the cell sap from parenchyma tissues. They inject toxins into the plant sap as a result, leaves become curly, they turn yellow, dry and fall down, flowering bodies disfigured, petals and sepals become curly. Hence, market value of flowers is adversely affected. The pest secretes honey dew like substance on plant which cause sooty mould. Further, it interferes with photosynthetic activities of plant and finally affect growth and quality of flowers.

Control Measures

Preventive

1. Collection and destruction of infested plant parts along with pest stages.
2. Scraping the infected stems or plant parts with wooden knife.

Curative Control

1. Fumigation with HCN gas is useful. However, this method needs expertise and technical knowledge for practice.
2. Spray phosphamidon 0.02 per cent (or)
3. Spray DDVP 0.02 per cent

Biological Control

Lady bird beetles feed on scale nymphs.

Lindigaspis rossi (Masket)

Marks of Identification

The pest has appearance like reddish brown waxy scale. On the old stems of rose reddish brown bodies are seen attached. The pest can affect shoots also. Pox like spots are seen on stem. The nymphs and adults insert their beaks into the plant part and suck the cell sap which results in curlying of leaves and disfiguring of flowering bodies.

Control

1. The pest can be controlled by spraying the crop 500 ml of Rogor 30 EC or Endosulphan 35 EC in 500 lit of water/ha.
2. Scraping the infected surface with wooden knife will reduce the pest population on the crop.

Thrips

1. *Rhipophorothrips cruentatus* Hood
2. *Dendrothrips* sp.

R. cruentatus (Figure 3)

Distribution

Widely distributed in India. The pest is reported from Punjab, Haryana, Himachal Pradesh, Maharashtra etc.

Marks of Identification

Adults are slender bodied, 1.4 mm long, blackish brown with two pairs of narrow and fringed wings. Nymphs are reddish or yellowish brown. They are minute and fast moving creatures. Eggs are bean or kidney shaped.

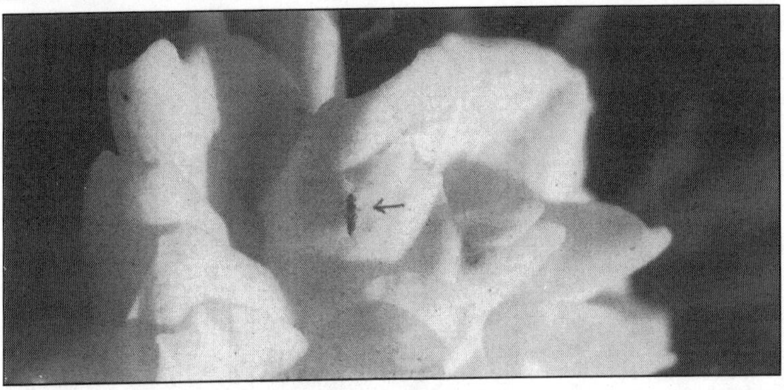

Figure 3: *Rhipiphorothrips cruentatus.*

Host Plants

This pest is highly polyphagus found attacking grapevine, jamun, rose, oak, *Calotropis, Procera,* guava, mango, almond, cashew nut etc.

Life Cycle

Both sexual and parthenogenetic reproduction occur side by side. Parthenogenetically males are produced by this insect pest. However, mated female of thrip lay eggs into the tender portion of leaves/buds/flowers/stems of rose by preparing slit. Kidney shaped eggs are embedded in plant tissue. The eggs get hatched within 3-8 days depending on climatic conditions. Newly hatched nymphs are tiny and reddish coloured. They scrape/rub their mouth parts over tender portion of rose and feed on oozing sap, as a result, white speaks are developed on leaves (Figure 4), the leaves become curly, flowers get disfigured and look unhealthy. The nymph mout for 3 times to become pupa and finally adult. Nymphal period is 11-22 days, pupal period is 2 to 5 days. Thus, within 14-33 days life cycle from egg to adult is completed. Males die after mating

Figure 4: Damage by Thrips.

within 2-6 days while, females survive for 18 to 20 days. About 5 to 8 generations are completed during a year.

Control Measures

Chemical control

Spray the crop with 0.03 per cent phasphamidon, or 0.03 per cent Diamethoate, or oxydemeton methyl.

Thrips florum Schmutz

Marks of Identification

Thrips are slender bodied insects measuring about 1.4 mm in body length. They are brown and provided with two pairs of wings which are fringed and narrow. Nymphs are still smaller than the adults and lacking wings.

Host Plants

Roses and other ornamental plants, grape vine, apple, Banana etc.

Life Cycle

As above species; Adult over winters in craks etc. Eggs are laid in February onwards on rose petals and on soft tissues. After hatching the egg, nymph moult for 3 times, become pupa and later adult.

Nature of Damage (Figure 5)

Flowers of rose are flecked with numerous light spots and streaks, which later darken and rot. Leaves may also show silver flecking.

Figure 5: Damage by Sucking Insects (Thrips).

Control Measures

As above.

Rose Caterpillars (Figure 6)

Caterpillar of the caster slug *Latoia lepida* Cram. and the caster hairy caterpillar *Euproctis lunata* Walker feed on leaves and flower petals. These caterpillars can be controlled by spraying the crop with 0.2 per cent phosphamidon or 0.03 per cent Azadirachtin and collecting and destructing caterpillars.

Rose Leafhoppers: *Edwardsiana rosae* (Jassidae)

Distribution

Europe, Asia.

Marks of Identification

Jassids are wedge shaped, small insects measuring about 2.00–2.5 mm in body length. They are greenish yellowish in colour and walk diagonally. Nymphs are light coloured, smaller in size, resembles with adults but lacking wings.

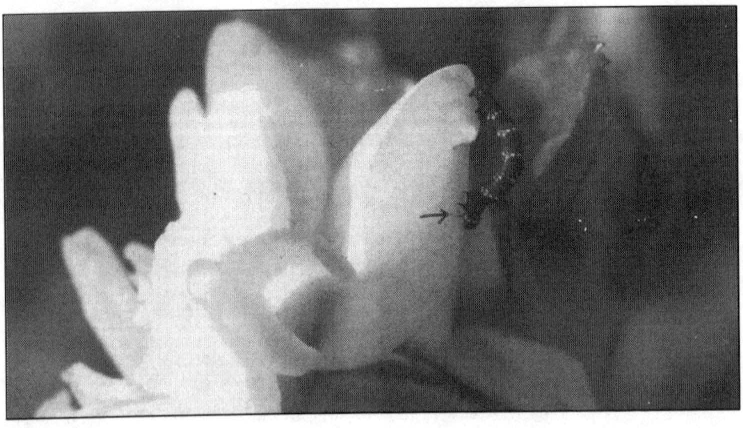

Figure 6: Rose Caterpillars.

Hosts

Roses.

Life Cycle

Eggs are embedded in tender parts of rose plant. After hatching the egg, newly emerged nymph start feeding on cell sap of rose by injecting its rostrum into tender parts. The nymph moult for 4 times to become an adult. Many generations are possible in a year.

Nature of Damage

Both nymphs and adults suck the cell sap from tender leaves, flowers, shoots and stem. While doing so, they inject toxins into the plant body as a result, leaves becomes curly, turn yellow, dry and drop down; flowers get disfigured by curling the petals. Thus, market value of flowers is adversely affected. They also secrete honeydew like substance on leaves that create sooty moulds which affect photosynthesis of the plant and further the growth and quality of flowers.

Control Measures

Spray the crop with 0.03 Azadirachtin or 0.02 per cent phosphamidon.

Rose Cushion Scale (Hemiptera) (Figure 7)

This sucking type of insect is mahogany reddish in colour but, it covers cottony white material. On the thoracic region orange brownish zigzag marking present. Nymphs are small sized.

Nature of Damage

Nymph and adult suck the cell sap from rose plant and affect the growth of plant and further quality and quantify of flowers.

Figure 7: Rose Cushion Scale.

Control Measures

Spray 0.02 per cent phophamidon or Azadirachtin 0.03 per cent.

Chapter 3
PESTS OF CHINA ROSE

China rose belongs to the family Malvaceae. It is native of China. Its common name is China Rose and vernacular names as Marathi–Jassvandi, Karnataka–Dasavala, Tamil Nadu–Mandaram, Bengali–Jaba etc. It has scientific name, *Hibiscus rosa-sinensis*.

China Rose is large plant of height 1-2 m. It is bushy shrub which provides numerous branches. Its leaves are broadly ovate, sharp toothed, green and ornamented. It is grown in sunny situation, porous soil, moderate temperature and relatively high humidity. On terminal on the branches, five petalled, large, single or double, rose-scarlet coloured flowers are developed. Flower petals are slightly arched back, and the attractive columns of pistils and stamens protrude out of the centre of the flower. There are many hybrid cultivars of *H. rosa-sinensis* with single or double flowers, are available for cultivation in India. Single flower cultivars refer to Agnes, Australian Rose, Lip stick, Netaji, My Beauty etc while, double flowers refer to Chitra, Mahatma, Daffodil, Bhaskar, Aurora, Alipore Beauty, etc. Indian hybrids released in the market for use are, Basant, Be-nazeer, Geetanjali, Narkati, Pakeezah etc. The flowers have medicinal value in curing baldness. This plant is widely grown in India but, attacked by several insect pests. Important pests of China rose are listed in Table 2.

The Cotton Aphid (*Aphis gossypii* Glover) (Figure 8)

Distribution

A. gossypii is world wide. It is well established polyphagous in tropics including several pacific islands. In temperate region it is major pest. In India, it has been reported from almost all states, specially Assam, Bihar, Punjab, Uttar Pradesh, Madhya Pradesh, Maharashtra, Andhra Pradesh, Tamil Nadu etc.

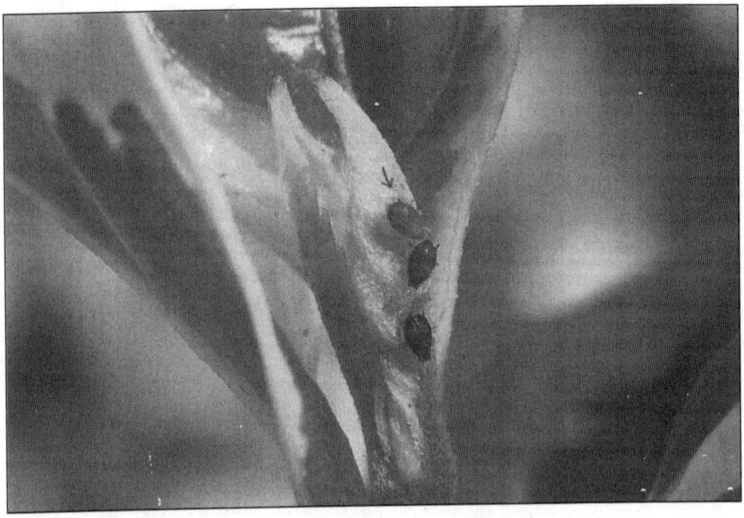

Figure 8: The Cotton Aphid (*Aphis gossypi* Glover).

Table 2: Insect Pests of China Rose.

Sl.No.	*Common Name*	*Scientific Name*	*Family*	*Order*
1.	The Cotton Aphid	*Aphis gossypii* Glover	Aphididae	Hemiptera
2.	The dusky cotton bug	*Oxycarenus laetus* Kirby	Lygaeidae	Hemiptera
3.	The milk weed bug	*Lygaeus civilies* Wolff.	Lygaeidae	Hemiptera
4.	The Red cotton bug	*Dysdercus koenigii* (Fab.)	Pyrrhocoridae	Hemiptera
5.	The caster caterpillar	*Euproctis lunata* Walker	Lymantridae	Lepidoptera
6.	Banded blister beetle	*Mylabris phalerata* Pallas	Meloidae	Coleoptera
7.	The mealy bugs	*Drosicha mangifera*	Coccidae	Hemiptera
		Pseudococcus citri Cock	Pseudococcidae	Hemiptera
		P. filamentossus Cock	Pseudococcidae	Hemiptera
8.	White fly	*Bemisia tabaci*	Aleyrodidae	Hemiptera
9.	Scale insect	*Aonidiella aurantii*	Diaspididae	Hemiptera
10.	Cotton weevil	*Myllocerus undecimpustulatus*	Curculionidae	Coleoptera

Marks of Identification

Aphids are soft bodied, louse like, greenish brown insects. Both winged and non winged forms are present. Winged forms are 1.1 mm to 1.8 mm long and non winged are 0.9 mm to 1.8 mm long. The colouration is mostly light green mottled with darker green and with siphunculi dark and a cauda pale or dusky. Nymphs resembling with adults except they don't show wings and are smaller in size.

Host Plants

A. gossypii is polyphagus pest. It is found attacking cotton, citrus, potato, okra, coffee, cucurbits, cocoa, cowpea, arhar, many ornamental plants including *Hibiscus rosa-sinensis* Linn., *Tecoma capensis* Lindl. *Rosa* spp., etc.

Life Cycle

Aphid reproduce parthenogenetically and viviparously giving birth to 10-20 nymphs per female per day. The nymph become adult within four moults. The pest complete its life cycle within 7-10 days. This species is anholocyclic, overwintering forms are seen in USA. Many generations are completed by this pest in India on agricultural and ornamental plants.

Nature of Damage

Both nymph and adult suck the cell sap from tender parts of China rose *viz.*, leaves, flowers, shoots, twigs, etc. resulting in curling of leaves & flowers. During March-April months its infestation is peaked on China rose and from September to October on Rose. The pest secretes honey dew like substance on leaves, that create sooty mould on leaves which interfere with photosynthetic activities of the plant affecting growth and quality of flowers adversely. The pest is also known as vector for viral transmission. The pest transmit over 50 plant viruses including lily rosette, lily symptomless and mosaic.

Control Measures

Preventive

1. Collection and destruction of infested plant parts along with pest stages.

Curative–Chemical

1. Spray the crop with Azadirachtin 0.03 per cent (or)
2. 0.03 malathion or 0.03 per cent Rogor.

However, the pest has developed the resistance against the orgonophophorus and carbamate insecticides.

Biological

Lady bird beetles like *Menochilus sexmaculus* feed on nymphs & adults of this pest and acts as good biocontrol agent.

Mealy Bug *Pseudococcus filamentosus* Cock. (Figures 9 and 10)

Distribution

India, Indonesia, Malayasia, China, USA etc.

Marks of Identification

There is sexual diamorphism. Males are winged and females are wingless, flat bodied with waxy filaments along the margins. The actual colour of female mealy bug is orange red but, its body is covered with mealy like cottony powder. Hence, the name mealy bug. Nymphs and females are with sucking type of mouth parts. The nymphs are flat bodied,

Figure 9: Mealy Bug *Pseudococcus* sp.

Figure 10: Mealy Bug *Pseudococcus* sp.

amber coloured and smaller than adults. Males are slender bodied with long antennae and wings.

Host Plants

China rose, citrus and many garden plants such as ferns, begonia, gardenia, poinsettia, cactus and other garden plants.

Life Cycle

About 300 or more eggs are laid in a single mass on the China rose. Eggs are with protective cottony mass. Eggs hatch within 10-20 days. Newly emerged nymphs crawl out and settle on suitable places (tender parts) for sucking the cell sap. The nymphs are covered with cottony waxy material. The nymphal period is about 6 to 8 weeks. The male nymphs pupate in cottony pupae and then emerged as winged adults from the pupae. All stages of mealy bugs are seen on the plant at the same time.

Nature of Damage

Both nymphs and adult females suck the cell sap from the tender parts of China rose including flowers, leaves, shoots and twigs etc. as a result leaves become curly and flowers get disfigured. The pest also secretes honeydew like substance which create sooty mould on leaves, ultimately affecting, photosynthesis, growth and yield of the crop adversely. Mealy bug colonies are seen on leaves, stems, buds and flowers.

Control Measures

Preventive

1. Collection and destruction of infested plant parts along with pest stages.

Curative

Spray the crop with 0.05 per cent carbophenothion or fenitrothion, (or) Malathion (or) 0.03 per cent phosphamidon. Other species mentioned in the list have also more or less same life cycle, nature of damage and control measures as given for above species. *Farrisia virgata* (Cock) (pseudococcidae) can also cause the damage to China rose which is characterized by two waxy whitish long filaments posteriorly and covered with mealy like waxy component.

Control

As above species.

Scale Insects (Figure 11)

As described under pests of rose.

Banded Blister Beetle (*Mylabris phalerata* Pallas)

Distribution

India: Maharashtra, Karnataka, Madhya Pradesh, Andhra Pradesh etc.

Marks of Identification

The beetle is with six alternating, bright orange and black bands on black background of the body. The beetle measures about 3 cm in body length. Grubs are whitish and found in soil. There is hypermetamorphasis in the larval stage.

Life Cycle

Round eggs are laid in soil. After hatching the eggs, newly emerged grubs feed on roots, humus and eggs of grasshoppers if any in the soil. Indefinite instars have been recorded in this pest. Pupation takes place in soil. The adults developed from pupae come out of soil and start feeding on flowering bodies of China rose.

Figure 11: Scale Insects on Stem

Host Plants

China rose, cucurbits, graminaceous grasses, *Ruetliia indica* Nees.

Nature of Damage

Grubs feed on roots and adults feed on flowers from July to September. Its population is peaked in August.

Control Measures

Preventive

1. Ploughing and digging the field for exposing (eggs, larvae and pupae) life stages of pest to natural mortality factors (biotic & abiotic).
2. Collection and destruction of adult beetles by keeping labour.

The Dusky Cotton Bug (*Oxycarenus laetus* Kby.) (Figure 12)

Distribution

South east Asia: India, Pakistan, Bangladesh. It is found throughout the India.

Marks of Identification

Adult bugs are 4-5 mm long, dark brown with dirty white transparent wings. The nymphs have rounded abdomen in early stage but in older instars the nymphs resembles with adult by shape but size is small and have no wings. Eggs are cigar shaped.

Host Plants

Cotton, Okra, holly hock, malvacous weeds, China rose, etc. The pest also feed on *Jasminum grandiflorum*, *J. multiflorum*, *J. humile* Linn., *Bauhinia acuminata* Linn., *Dambeya natalensis*, *Plumeria acuminate* (Linn.) *Bougainvillea* spp., etc.

Life Cycle

Mated female lay eggs in the spring on China rose. Eggs are whitish but turn light pink at the time of hatching. Eggs are laid singly or in small cluster. Each cluster may contain

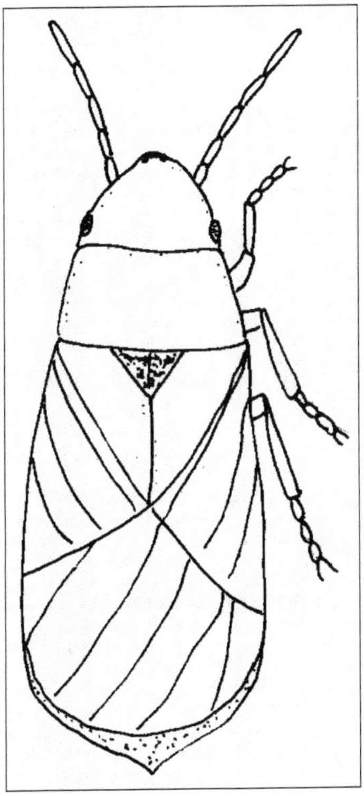

Figure 12: The Dusky Cotton Bug.

5 to 20 eggs. They hatch in about 5-10 days. Nymph feed on tender parts of China rose by sucking cell sap and moults for seven times and become an adult within 30-40 days. Many generations are completed by this pests during a single year.

Nature of Damage

Both nymphs and adults suck the cell sap from tender parts of China rose and affect growth of plant and quality and quantity of flowers. The individuals mostly feed on terminal portion of the plant.

Control Measures

Preventive

1. Collection and destruction of eggs, nymphs and adults of the pest.

Curative

1. Spray the crop with 0.04 per cent monocrotophos (or)
2. Dusting the crop with 5-10 per cent DDT/BHC.

Biological Control

Triphleps tantilus Mot. (Hemiptera) feeds on nymphs of the pest.

Red Cotton Bug (*Dysdercus koenigii* Fab.) (Figure 13)

Distribution

Punjab, Maharashtra, Uttar Pradesh, Karnataka, Andhra Pradesh etc.

Marks of Identification

Adult bug is elongated, slender bodied, crimson red with white transverse bands on abdomen. Scutellum, antennae and membraous portion of fore wings are black. Black spot on the wing. Eggs are cigar shaped. Nymphs are adult like but smaller in size and lacking the wings.

Host Plants

Jasminum sambae Linn., (April-December), *Dombeya spectabilis* Cav., *D. natalensis* Sind; *Carissa carandas* Linn., *Budleia madagascarensis* Lamk., Cotton, China rose etc.

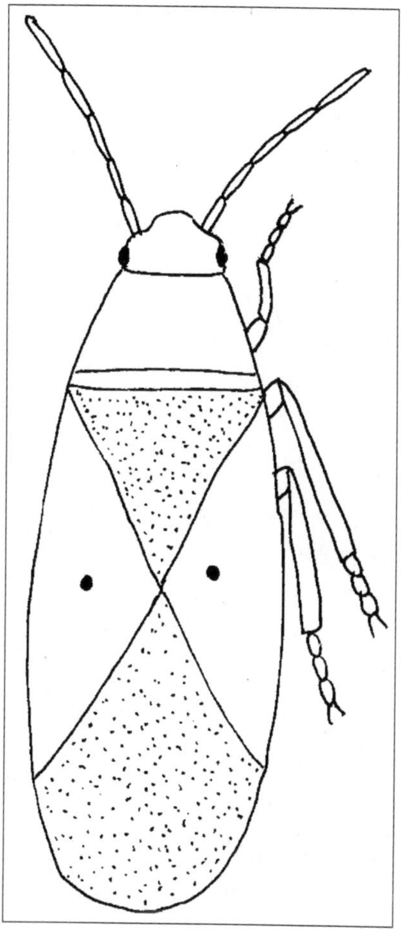

Figure 13: Red Cotton Bug.

Life Cycle

A single mated female can lay about 100-130 eggs in moist soil or in crevices of ground. Eggs are laid in clusters. Each cluster may contain 70-80 eggs. Incubation period is 7-8 days. Nymphal period is 32-38 days. Thus, life cycle from egg to adult is completed with in 37 to 43 days. The adult can survive for 3 months in winter.

Nature of Damage

Both nymphs and adults suck the cell sap from the tender parts of the China rose due to which leaves becomes curly, turn yellow, becomes dry, flowers get disfigured affecting quality. Flowering capacity is adversely affected by sucking the cell sap due to the pest.

Control Measures

Preventive

1. Collection and destruction of eggs, nymphs and adults from the China rose.
2. Infected plant parts should be collected and destroyed.

Curative

1. *Chemical*: As above pest.
2. *Biological control*:
   (*a*) *Antilochus cocqueberti* F. (Pyrrhocoridae) and
   (*b*) *Harpactor costalis* Rev. (Reduviidae) feed on nymphs of the pest.

The Cotton White Fly (*Bemisia tabaci* Gennadius)

Distribution

Maharashtra, Bihar, Madhya Pradesh, Punjab, Haryana, Rajasthan, Karnataka etc.

Marks of Identification

Adults are small moth like creatures, whitish in external body colouration with transparent waxy wings and 1.0 to 1.5 mm in body length. White wings are in two pairs. Nymphs are louse like, pale yellowish, sluggish, clustering together on under surface of China rose leaves. Nymphs are elliptical. Eggs are stalked and subelliptical.

Host Plants

Lantana camera Linn., *Euphorbia pulcherina* Wild.; *Dombeya spectabilis* Cav.; *Poinciana pulcherima* (Linn.), Cotton cabbage, cauliflower, Okra, Bajra, Melon, Mustard, China rose, Toria, Potato, Brinjal, some weeds.

Life Cycle

The mated female lay eggs singly on the under surface of the leaves of China rose. On an average 120 eggs are laid per female. Incubation takes place within 3-5 days in April-September and 7-15 days in October-November and 33 days in December–January. The nymphal period is 9-14 days in April-September and 17-80 days in October-March. Pupal period is about 2-8 days. The life cycle from egg to adult is completed within 15-120 days. As many as 11 generations are completed in a single year.

Nature of Damage

Both nymphs and adults suck the cell sap from leaves, buds and flowers and affect the quality of flowers by disfiguring. The pest is also responsible for forming sooty mould on leaves and affecting growth and yield of the flowers.

Control Measures

Preventive

1. Collection and destruction of eggs, nymphs, pupae and adults of pests alongwith infested plant parts.

Chemical

1. Spray 0.1 per cent malathion (or)

2. Spray 0.025 per cent methyl demeton
3. Spray 0.01 per cent phosphamidon
4. Spray 0.02 per cent Endrin

Biological control

By using

1. *Chrysoperla* sp. (Neuroptera)
2. *Brumus* sp. (Coleoptera).

Chapter 4
PESTS OF JASMINE
(Chameli/Janti)
(*Jasminum* spp.)

The Jasmine (*Jasminum* spp.) produces scented white flowers. Hence used in production of scent fumes. Preparation of garlands for sale has also great potential. According to Rendle (1925) there are 200 species of the genus *Jasminum*. However, Veluswamy *et al.* (1973) revised the species and declared only 89 species. Out of the species with scented flowers only three are commercially cultivated for their cut flowers and for extraction of perfume which refer to *J. auriculatum, J. grandiflorum* and *J. sambae*. Other species cultivated in India as ornamental value refer to *J. humile, J. arborescens* and *J. pubscens* var. *rubescens* (China). *J. grandiflorum* is native of India and known by "Royal Spanish", "Catalonian" or Italian Jasmine. In India, it is known as "Chameli". It is characterized by leaves shining dark green with 5-7 leaflets. Flowers are white with a purplish tinge and highly scented. *J. sambae* has highly scented flowers borne in clusters of 3-12. Flowers may be single/semi double/perfectly double. *J. arborescens* have blooming mainly during winter months (nabamallika). *J. humile* shows 1-3 paired leaflets and yellow faintly scented flowers.

The Jasmine is attacked by a large number of pests. Important pests with their taxonomic position are tabulated in Table 3.

The Jasmine Thrip (*Thrips orientalis* Bangnall.)

Distribution

India.

Marks of Identification

Thrips are slender bodied, brownish insects which contain two pairs of narrow and fringed wings. Adults measures about 1.00 mm in body length. Nymphs are still smaller than adult and lacking wings. Eggs are kidney shaped or bean shaped.

Host Plants

Jasmine *Jasminum multiflorum, J. grandiflorum.*

Life Cycle

Eggs are laid into tender tissues of the Jasmine. After hatching the eggs, newly emerged nymphs feed on tender buds, leaves and flowers of the jasmine and create white specks on leaves, and yellowish specks on petals of the flower. The full grown nymph moult into pupa which do not feed. The pupa is finally transformed into adult thrip. Many generations are possible in a single year.

Table 3: Pests of Jasmine.

Sl.No.	*Common Name*	*Scientific Name*	*Family*	*Order*
1.	The Jasmine Thrip	*Thrips orientalis* Bangnall	Thripidae	Thysanoptera
2.	The Jasmine Bud worm	*Nausinoe geometralis* (Guen.)	Pyraustidae	Lepidoptera
3.	The Red cotton bug	*Dysdercus koenigii* Fab.	Pyrrhocoridae	Hemiptera
4.	The Eriophyid mite	*Aceria jasmini* Chanana	Eriophyidae	Acarina

Nature of Damage

Both nymph and adults feed on tender shoots, tender leaves and petals of flower by rubbing their mouth parts over tender portions and feeding on oozing sap of the crop plant. Appearance of the plant and flowers adversely affected due to such feeding by the pest. The insect cause decaying look to crop.

Control Measures

Spray the crop with 0.03 per cent Azadirachtin or 0.03 per cent Malathion (or) 0.02 per cent Phosphamidon

The Jasmine Budworm [*Nausinoe geometralis* (Pyraustidae)]

Distribution

Australia, China, Myanmar, Formosa, Java, Sri Lanka, India, West Africa, Pakistan.

Marks of Identification

The young worms are light yellow but older ones become darker. The pupa is obtect type, brown and tapering towards posteriorly. Moth is brown with hyaline patches on the wings. The abdomen of the moth contains lateral patches of lighter shades on each segment. The abdomen is purplish brown in colour. Eggs are greenish yellow and rounded.

Host Plants

Jasmine.

Life Cycle

A mated female can lay about 15-20 eggs singly on leaf lamina of Jasmine plant. Eggs hatch in about 3-4 days. The larval period is 12 to 15 days. There are five instars in the larval forms. The full grown larvae spin webbings in the shaded portion of the plant along with leaves and pupate in the typical shelter prepared. Pupal period is 6 to 7 days. The life cycle is completed within 22 to 24 days. The pest is mostly found from July to August.

Nature of Damage

The caterpillars feed on leaves and in severe infestation they can skeletonize the entire plant. During rainy season caterpillars feed on leaves of shaded portions and during dry season they feed on terminal portion of the plant. The caterpillar make webbings in the leaves and shelter themselves. The caterpillars in first few instars are gregarious in condition. As a result of collective attack by gregarious caterpillars, the vitality of crop is affected. Flower setting and quality and quantity of flowers is adversely affected.

Control Measures

Preventive

1. Collection and destruction of caterpillars and pupae from webbing of plant.

Chemical Control

1. Spray the crop with 0.03 per cent Azadirachtin (or)

   0.03 per cent Malathion (or)

   0.02 per cent Phosphamidon.

Red Cotton Bug (*Dysdurcus koenigii*) (Figure 13)

Details are discussed under pests of china rose.

The Eriophyid Mite

Suck the plant sap.

Chapter 5
PESTS OF CHRYSANTHEMUM

Chrysanthemum is commonly called as "Glory of the East" or as "Mum" in short in USA. Chrysanthemum is widely cultivated in the world as garden flowers. It ranks next to the rose in popularity. It has very wide range of flower shades. The florist's chrysanthemum is a short day plant. For proper vegetative growth plant requires longer time. Chrysanthemums were cultivated in China at least 3,000 years ago. *Crysanthemum morifolium* and C. *indicum* produced several hybrid varieties in the world. The important species of *Crysanthemum* are given below:

1. *Crysanthemum morifolium*
2. C. *indicum*
3. C. *ornatum*
4. C. *hortorum*

Crysanthemum flowers have good export potential. Hence, the above species of *Crysanthemum* plays important role in foreign exchange. Crysanthemum floriculum is attacked by several insect pests. Important amongst them are listed in Table 4.

The Sunflower Lacewing Bug (*Cadmilos retiaris* Dist.) (Tengidae: Hemiptera)

Distribution

Widely distributed in India where sunflower and chrysanthemum are grown.

Marks of Identification

The adult lace wing bug is typically reticulated with lace like wings. The wings are transparent. The adult measures about 4 mm in body length and are blackish in body colour. Nymphs are lighter than adults in body colouration but, they have also reticulation on their bodies.

Host Plants

Chrysanthemum, sunflower, Gaillardia, Marigold, Launea, Vernonia, daisy, Aster, Scabiosa, weeds like *Argemone* spp. etc.

Life Cycle

Eggs are laid on Chrysanthemum on the upper surface. Eggs are laid singly in plant tissue. They are placed obliquely in the tissues leaving the opercula exposed which seen as whitish or brown dots. Incubation period is 5-7 weeks. Nymphal period is 2 to 3 weeks during which the nymph moults for five times to become adult. For begin with first generation of new season, the adults appear in July. It is found quite active upto September in breeding and

Table 4: List of pests on Chrysanthemum.

Sl.No.	*Common Name*	*Scientific Name*	*Family*	*Order*
1.	Chrysanthemum lace bug	*Corythuca marmorata*	Thripidae	Hemiptera
		Cadmilos retiaris	Thripidae	Hemiptera
2.	Chrysanthemum mealy bug	*Phenacoccus gossypii*	Pseudoccidae	Hemiptera
		Pseudococcus filamentosus	Pseudoccidae	Hemiptera
3.	Thrips	*Thrips* sp.	Thripidae	Thysanoptera
4.	White fly	*Bemicia* sp.	Aleurodidae	Hemiptera
5.	Variagated cutworm	*Peridroma saucia*	Noctuidae	Lepidoptera
6.	White flies	*Aleurodicus dispersus*	Aleurodidae	Hemiptera

damaging various host plants mentioned above. Many generations are completed in a single year.

Nature of Damage

Both nymphs and adults suck the cell sap from the tender parts of chrysanthemum plant and inject toxins into the plant body resulting curlying of leaves and petals of the flowers. The infested leaves turn yellowish brown, their edges start drying, finally the entire leaf is dried and drop down. Flowers become disfigured, colour of flowers get changed, they don't look fresh and healthy. Therefore, market value of flowers is adversely affected. Leaves get typically bleached. The pest can also injure stems.

Control Measures

Preventive

1. Collection and destruction of infected plant parts along with pest stages.

Curative

1. Spray the crop with 0.05 per cent Rogor (or)
2. 0.03 per cent DDVP or 0.03 per cent phosphamidon (or)
3. 0.03 per cent Malathion/diazinon or sevin.

Mealy Bugs

Several species of mealy bugs are associated with chrysanthemum which includes citrus mealy bug. *Pseudococcus filamentosus, P. citriculus*, etc.

P. citriculus

Distribution

India. It is reported from both, South and North parts of India. It is more common in southern India.

Marks of Identification

Adult female is flat bodied, wingless with short waxy filaments along the margins. Male is winged with long antennae and large wings. However, the male do not have mouth parts. Hence, male do not feed and do not cause any damage to our crop. Nymphs are flat bodied, smaller than adults, amber coloured with waxy mealy like powder on their back and filaments.

Host Plants

Citrus, Begonia, Gardenia, Poinsettia, Cactus, many green house plants, several floricultural plants.

Life Cycle

The eggs are laid in protective mass on chrysanthemum plant. Incubation period is 10-20 days. Nymphal period is 6 to 8 weeks. Male pupate in cotton like cocoon for 2 to 3 weeks and become adult. Thus, the life cycle in females is completed within 8 to 11 weeks. Several overlapping generations are common. Hence, all stages of pest are seen simultaneously on the plant.

Nature of Damage

Males are not destructive. Only females and nymphs of both sexes are destructive to crop. They suck the cell sap from tender parts of the plant and affect the growth of plant adversely including quality and quantity of flowers. The pest also create sooty mould on the plant and affect the flower quality.

Control Measures

Preventive

1. Collection and destruction of plant parts infected by the pests along with all pest stages.

Curative

Chemical Control

1. Spray the crop with 0.02 per cent diazinon or dursban (or)
2. Spray the crop with 0.03 per cent Malathion (or) 0.03 per cent Azadirachtin.

Biological Control

Lady bird beetles and lace wings feed on nymphal forms of this pest.

Thrips

1. *Thrips tabaci*
2. *Rhipiphorothrips* sp.

Distribution

India.

Marks of Identification

Thrips are measuring about 1 mm in body length. They are with narrow and fringed wings and brownish body colouration. Antennae and legs are short. Nymphs are still smaller than their adults.

Host Plants

Chrysanthemum sp. and other green house plants.

Life Cycle

Bean shaped eggs are laid in the plant tissue singly. After hatching the eggs, newly emerged nymphs start sucking the cell sap of tender portion of the plant by rubbing mouth parts upon them and moulted and pupated. Pupae are as an abnormal case of incomplete metamorphic order and pupa finally become adult.

Nature of Damage

Both nymph and adult rub their mouth parts on tender portion of the crop and feed on oozing sap. As a result of tearing and rubbing the plant portion, white or yellow specks are seen on the flowers and leaves which affect the photosynthesis, growth and yield of the flowers.

Control Measures

Chemical

1. Spray the crop with 0.03 per cent Malathion (or)
   0.03 per cent Azadirachtin (or)
   0.02 per cent Dichlorvos (or)
   0.02 per cent Phosphamidon or sevin.

Variagated Cutworms

Distribution

India, U.K., Europe, Asia.

Marks of Identification

Destructive stage is caterpillar. It is with 3 pairs of walking legs and peudolegs on abdomen. Adult is moth.

Life Cycle

Eggs are laid on the plant by the moth female. After hatching the eggs, larvae feed on the leaves of *Chrysanthemum* of outdoor plants and skeletonize them in severe infestation. Caterpillar moult for 5 times to become pupa which is brown and obtect type. Pupal period is 6-12 days. The adults are moths. The moth lay eggs on the leaves of chrysanthemum and again the life cycle is progressed in the typical manner on the crop.

Nature of Damage

Only caterpillars damage the crop by feeding upon leaves. In severe infestation, entire plant is skeletonized by the pest.

Control Measures

Preventive

1. Collection and destruction of caterpillars found on the crop.

Chemical

1. Spray the crop with
   0.03 per cent Malathion or
   0.03 per cent Azadirachtin or
   0.05 per cent Endosulphan

Whitefly

1. *Bemisia tabaci*
2. *Aleurodicus dispersus* (Figure 14)

Distribution

Both species are widely distributed in oriental region.

Marks of Identification

Adult look like small moth. It is clear winged with two dark spots. Nymph is with short glass like roods of wax along the sides of body. Adults always fly at evening and morning.

Host Plants

Polyphagous; several horticultural, vegetable and floricultural plants are affected by this pest.

Figure 14: Whitefly *Aleurodicus dispersus.*

Life Cycle

Eggs are laid on crops. Incubation period is 2-5 days, Nymphal period is 14-20 days, pupal period is 2-5 days. Within 21-31 days life cycle is completed. Many generations are possible in a single year.

Control Measures

Preventive

1. Infected plant parts should be collected and destroyed along with pest stages.

Chemical

1. Spray the crop with

   0.03 per cent Malathion

   0.03 per cent Azadirachtin

   0.02 per cent Phosphamidon.

Chapter 6
PESTS OF NERIUM

Nerium belongs to the family Apocynaceae. Its common names are sweet scented oleander and Rose Bay. In Hindi it is referred as Kaner or Kanel. It is one of most popularly grown shrubs in India. It is widely cultivated from northern part of India than southern part.

Nerium in general is characterized as a spreading bushy shrub. It gives number of cane like stems from ground level. The most common colours of flowers are white and pink. Deep red, cream and shades of pink are also available in Indian forms. Nerium attains a height of 2.5 to 4.00 m. It has linear and lanceolate leaves. The flowers are given out in large terminal cymes, a typical bunch of flowers is seen at the tip of the branch. The blooming period is from April. The flowers have a characteristic sweet scent. In certain parts of India, West Bengal, pruning is done for better harvest of leaves. The old growth are removed after new growths. The plant is propogated by cuttings or by layering. However, the white juice of the plant is poisonous and should not be given to cattle for eating purpose even flowers can cause death if eaten.

Flowers of Nerium are very good looking, attractive, milled scented and resembles with rose flowers. Hence, they have tremendous demand in preparing garlands in India and have tremendous export potential in floriculture business.

There are several species of Nerium known to science. But, 3 species are commonly cultivated in India which refer to:

1. *Nerium oleander*
2. *Nerium oleander* var. *flore pleno*
3. *Nerium odorum* (syn., *N. indicum*)
4. *Nerium oleader* var. *variegatum* (creat white leaves)

The species *N. oleander flore plano* has Mediterranean origin while, *N. oleander odorum* has origin from North India and Japan.

The Nerium is attacked by several insect pests. Important amongst them are listed in Table 5.

The Cottony Cushion Scale (*Icerya purchasi* Maskell) (Coccidae-Hemiptera)

Distribution

Tropical and subtropical regions of the world. It is native of Australia. From India, it has been widely recorded on wild plants from southern region.

Table 5: Pests of Nerium.

Sl.No.	Common Name	Scientific Name	Family	Order
1.	Jasmine thrip	*Thrips orientalis*	Thripidae	Thysanoptera
2.	Aphids	*Aphis* spp.	Aphididae	Hemiptera
3.	Oleander scale	*Aspidoitus nerii*	Coccidae	Hemiptera
4.	Cottony cushion scale	*Icerya purchasi*	Coccidae	Hemiptera
5.	Hemispherical scale	*Saissetia hemispherica*	Coccidae	Hemiptera
6.	Mealy bug	*Pseudococcus longispinus*	Pseudococcidae	Hemiptera
7.	Termites	*Reticulitermes* sp.	Termitidae	Isoptera

Hosts

Nerium, *Acacia* spp., guava, pomegranate, fig, grapevine, peach, walnut, apricot, almond, apple, wattle, casurina, gorse etc.

Marks of Identification

Adult female is reddish brown, soft-bodied with flat and oval shape. Female has fluted egg sac which is quite large and whitish in colour. The female is marked by ridges and by white cottony substance with which large number of eggs are found. Nymphs are reddish brown to brick red, oval, 3.0 mm in length and 1.5 mm in width. Males are rare.

Life Cycle

The pest can reproduce parthenogenetically. However, a single female lays about 700 eggs in the ovisac of the body. Incubation period is 24 hours. Newly emerged nymphs are reddish in colour. The incubation period is delayed for several weeks in cold weather. The first instar nymphs are crawlers. They search for suitable spot for settlement on the plant for sucking the cell sap. The nymph moult for three times to become adult. Within 46-240 days life cycle is completed. The duration is dependant on climatic conditions. In colder weather duration of life cycle is delayed. The pest is active throughout the year. However, in summer it multiplies very fastly.

Nature of Damage

Both nymphs and adults suck the cell sap from Nerium plant and affect the growth of plant and finally the quality and quantity of flowers. Leaves turn yellow, they dry and drop down. In severe infestation, the entire plant is covered with the scales and plant dries and dies.

Control Measures

Preventive

1. Keep watching the garden continuously, if pest infestation noticed, remove the infected plant parts along with pest stages and destroy.

Curative

Chemical Control

1. Spray the following insecticides on the crop–

   0.03 per cent Malathion (or)

   0.03 per cent Sevin.

Hemispherical scale (*Saissetia hemispherica*) (Coccidae–Hemiptera)

Distribution

India, Europe, U.K., U.S.A. etc.

Marks of Identification

The mature scales are highly convex, hemispherical, smooth, brown in colour and about 2 mm in body length. Nymphs are flat bodied and oval.

Host Plants

Nerium (Oleander).

Life Cycle

The mature female can lay about 500 eggs beneath her body. The first instar nymphs are crawlers. They select the suitable place for sucking the cell sap from leaves or stems of Nerium. The nymph moult for 3 times and become an adult within 4 to 6 months. Thus, within 6 months one generation is completed by this pest.

Nature of Damage

The nymphs feed along the veins of the leaves and are active for a month or more. Both adults and nymphs suck the cell sap from leaves, flowers and stems and affect the growth of plant and quality and quantity of flowers.

Control Measures

Preventive

1. Collection and destruction of infested plant parts.
2. Scrapping the infected leaves and stems with wooden knife for removing the scales.

Chemical Control

Spray the crop with 0.03 per cent Malathion and 0.03 per cent sevin alternatively and repeatedly.

White Scales (Figure 15)

Suck the cell sap of crop.

Jasmine Thrips (*Thrips orientalis*)

Pl. see under pests of Jasmine.

Aphids (*Aphis* spp.)

Pl. see under pests of rose and China rose.

Mealy Bugs

1. *Pseudococcus longispinus* (Pseudococcidae–Hemiptera) (Figure 16)
2. *Planococcus citri*

Distribution

India, Europe, U.S.A.

Figure 15: White Scales.

Marks of Identification

Mealy bugs are flat bodied in adult female stage and in nymphal stages. Adult males are winged. There is sexual diamorphism in this pest.

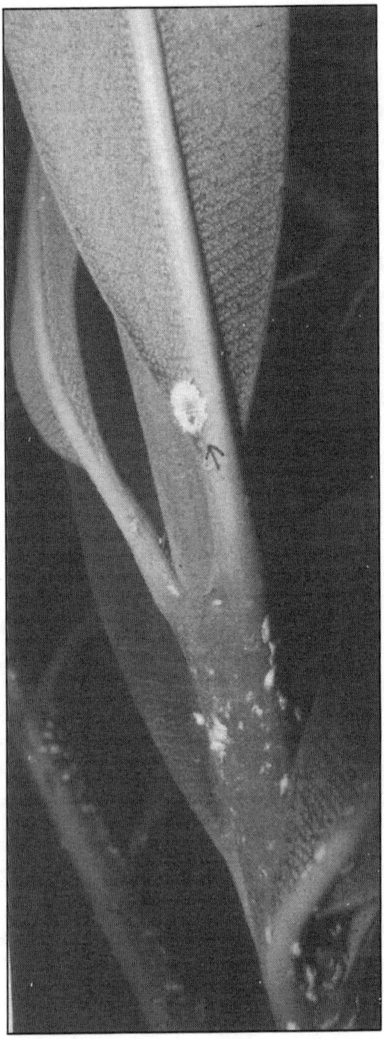

Figure 16: Mealy Bugs.

Host Plants

Nerium spp., citrus, etc.

Life Cycle

Eggs are laid on Nerium plant. The nymph feed on tender portion of Nerium and moult into pupa in case of male and then into adults. Nymphal period is about 4 to 7 weeks depending upon climatic conditions. All stages of the pest are seen on crop simultaneously.

Nature of Damage

Both female adults and nymphs suck the cell sap from the Nerium plant and affect the quality and quantity of flowers.

Control Measures

Chemical

Spray the crop with 0.03 per cent Malathion (or) 0.02 per cent Phosphamidon

Biological

Lady bird beetles and lace wings feed on mealy bugs.

Termites (*Reticulitermes* sp.) Termitidae–Isoptera

Distribution

Europe, Asia.

Marks of Identification

Termites are social insects. There is caste system, division of labour and polymorphism in termite. They can digest cellulose and look similar to ants as their bodies are whitish (Abdomen + thorax) hence they are called as white ants.

Host Plants

Polyphagous: *Nerium* sp., several agricultural & forest plants.

Life Cycle

Female queen is machine of egg laying. Queen lays about 80,000 eggs in 24 hours. Incubation period is 1 to 6 weeks. The reproductive castes when produced required 1-2 years for maturation. Many generations are completed during a year.

Nature of Damage

Termites damage Nerium by constructing their nest upon the trunk. They feed on bark and wood and kill the plant.

Control Measures

Preventive

1. Destruction of termetaria.

Chemical

1. Spray termetaria with Aldrex and Termex.
2. Treat infected surface with chloropyriphos.

Chapter 7

PESTS OF MARIGOLD

Marigold (Figure 17) belongs to the family Compositae. It has African origin. Marigold is grown in different soil and climatic conditions. It has very attractive flower colours and bloom for a considerable long period. The flowers can be preserved in drying condition for long period. Marigold is very good annual flower of India. It has great importance for garden display and commercial cultivation. It is quite useful for preparing garlands. Cut flowers and even whole plant can be used for decoration. They are planned for bed mass display and mixed borders. Flowers of marigold are in different colours and shades. Marigolds are classified into two groups namely,

1. French Marigold (*Tagetes patulo*)
2. African Marigold (*Tagetes erecta*)

Figure 17: Marigold .

French Marigold

These plants are 30-37.5 cm tall, flowers may be in singles or doubles. The colour range is deep scarlet, manogany and rusty reds, primrose, golden yellow, orange, yellow and with combinations. These marigolds are compact in habit. "Ursula", "Golden Gem" and Lulu are from single–single dwarf (30-37 cm). Red and Gold hybrids are produced by crossing French and African marigolds. French marigolds start flowering within 2 to 2.5 months.

African Marigold

These plants are vigorous tall, growing upto 90 cm in height and produce large globular flowers, about 15 cm or more across and are in shades of Lemon yellow, Golden yellow, Bright yellow, orange or whitish. From USA a white cultivar is released recently. Dwarf types (20-30 cm) with similar flower heads can also available for use. Carnation–flowered marigold is most beautiful creation from this group. "Apricot", "Sun Giants", Guinea gold, "Fiesta", "Golden yellow, Primrose etc. are important. The African groups can start flowering within 2.5 to 3 months.

Marigold plants are damaged by several insect pests. Important pests are listed in Table 6.

Japanese Beetle (*Popillia complanata*)

P. cyanea Hope

Distribution Japan, U.S.A.

Marks of Identification

The beetles are blackish in colour. Grubs are whitish with head brownish-dark and three pairs of walking legs.

Host plants : Marigold

Life cycle

Eggs are laid in soil. The grubs feed on roots of marigold and various other plants in soil and with humus material.

Table 6: Pests of Marigold.

Sl.No.	Common Name	Scientific Name	Family	Order
1.	The sunflower Lacewing bug	*Cadmilos retiaris* Dist.	Tingidae	Hemiptera
2.	The thrips	*Thrips tabaci*	Thripidae	Thysanoptera
3.	Japanese beetle	*Popillia japonica*	Melolonthidae	Coleoptera
4.	Tarnished plant bug	*Lygus lineolaris*	Lygaeidae	Hemiptera
5.	Leaf hoppers	*Empoasca fabae*	Cicadellidae	Hemiptera
		E. devestans	Cicadellidae	Hemiptera
6.	Green house leaf Tier	*Udea rubigalis*	-	Lepidoptera
7.	Blister beetle	*Zonabria postulaa*	Meloidae	Coleoptera
8.	Gram pod borer	*Helicoverpa armigera* (Hubn.)	Noctuidae	Lepidoptera

Pupation takes place in soil, Adults, thus formed in soil emerge out of soil at night and start feeding on marigold leaves and flowers.

Nature of Damage (Figures 18 and 19)

Beetles feed on flowers and leaves of marigold and destroy flowers while, grubs feed on roots of marigold.

Control Measures

Preventive

1. Collection and destruction of beetles.
2. Ploughing or digging the field for exposing the immature stages of the pest to natural mortality factors.

Chemical Control

1. Spraying–0.04 per cent diazinon/Fenitrothion
2. Dusting 10 per cent BHC 20 Kg/ha.

Biological Control

1. *Bacillus popilliae* Dutkey cause mortality in grubs of Japanese beetle.
2. A mixture of 9 parts of geraniol and 1 part of eugenol serves as a food lure for Japanese beetle. This can be used for attracting beetles and then collecting and destroying.

White Grub Beetle (*Holotrichia consanguinea*)

Damage and control as above.

Marigold Thrips (Figure 18)

Distribution

India.

Figure 18: Damage by Thrips.

Marks of Identification

Thrips are small insects measuring about 1 mm in body length. Slender bodied insects shows two pairs of narrow

Figure 19: Damage by Thrips.

wings which are fringed. Nymphs resemble with adults but they are miniatures and lacking wings.

Host Plants

Marigold, vegetables, ornamental flowers etc.

Life Cycle

Eggs are bean shaped. They are inserted into plant tissue by ovipositor. Newly hatched nymphs and older nymphs feed on the oozing sap of the marigold when they rub their mouth parts on tender portion of leaves and flowers. The full grown nymph moult into pupa and finally the adult.

Nature of Damage

Both nymphs and adults scrape their mouth parts on tender portion of the plant and feed on oozing cell sap as a result, white specks are developed on leaves (Figure 19) and on flowers (Figure 18) affecting quality of flowers.

Control Measures

1. Spray the crop with 0.03 per cent Malathion (or)
   0.03 per cent Azadiractin (or)
   0.02 per cent Phosphamidon

Leaf Hoppers (*Empoasca devestans* Dist.) (Figure 20)

Distribution

U.S.A., South East Asia.

Marks of Identification

E. devestans

Adults are about 3 mm in body length; greenish yellow and walk diagonally. Nymphs are similar to adults except they are miniatures and lacking wings.

Host Plants

Marigold, cotton, vegetables etc.

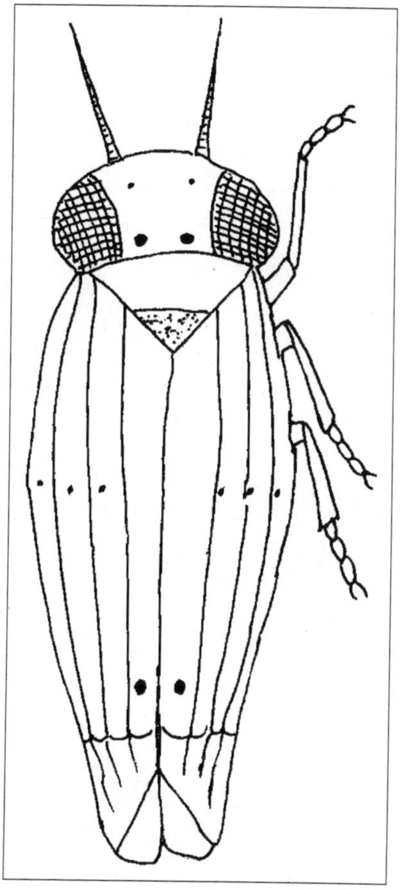

Figure 20: Jassid (*Empoasca devestans* Dist.).

Life Cycle

Eggs are embedded into marigold tissues. Incubation period is 4-11 days. Nymphal period is 7 to 21 days. The life cycle is completed in 11 to 32 days depending upon climatic conditions. Many generations are completed during a year subject to condition of availability of host plants.

Nature of Damage

Both nymphs and adults suck the cell sap from tender leaves, shoots, stems and flowers and affect the growth and yield of flowers. The pest also secrete honey dew like substance on leaves which create sooty mould. The sooty mould interfere with photosynthetic activities of the plant that affect the growth and yield of the crop adversely.

Control Measures

Chemical

1. Spray the crop with 0.02 per cent phosphamidon (or)

   0.03 per cent Malathion/Methyldemeton/Rogor.

Blister Beetle (*Zonabria postulaa*)

Distribution

India.

Marks of Identification

Beetles are black in colour with yellow strips across their wings (elytra). The grubs are whitish and found in soil.

Host Plants

Bajra, jowar, maize, marigold, china rose, grasses, cucurbits, etc.

Life Cycle

Eggs are whitish and rounded. Female lay eggs in soil. Incubation period is 15 days. Larval period is not known since there is hypermetamorphosis in the larvae (grubs). Pupation takes place in the soil. The beetles emerged from pupae from soil remain active for five months only *i.e.* August to December.

Nature of Damage

Beetles feed on flowers of marigold and the grubs feed on roots of marigold and other plants.

Control Measures

Preventive

1. Ploughing and digging the field for exposing immature stages of the pest to natural mortality factors (biotic and abiotic).
2. Collection and destruction of beetles.

Mechanical

Beetles get attracted to light hence, use light traps for catching them.

Chemical

Dusting 10 per cent BHC at 20 Kg/ha.

Defoliating Caterpillars

Some lepidopterous caterpillars (*Spilosoma obliqua*, *Spodoptera* spp., etc) feed on leaves and flowers of marigold. Therefore, they should be controlled by spraying the crop with following insecticides.

1. 0.03 per cent Malathion (or)
2. 0.03 per cent Azadirachtin
3. 0.03 per cent Phosphamidon.

Mealy Bug (Figure 21)

Suck cell sap. Control given in earlier Chapter 5.

Figure 21: Mealy Bug.

Leaf Miner (Figure 22)

Figure 22: Damage by Leaf Miner.

Chapter 8

PESTS OF NIGHT JASMINE (Parijatak)

Night Jasmine belongs to the family Oleaceae. It is native of India. In Hindi it is referred as “Harisingar”, in Bengali–"Seuli”, in Kannada “Parijata” and in Marathi–Parijatak. Scientific name of this plant is *Nyctanthes arbor-tristis*. This plant is about 4 m tall and is bushy shrub with rough hairy leaves. It can grown in any garden soil without much care. Night Jasmine contains sweet scented flowers with white corolla and an orange red tube and centre. Autumn and winter are the flowering seasons. The very special feature of the flower is that the flowers open at night and start failing at day break. Pruning is done after each flowering from long woody branches. This plant can be propogated by cutting during the rains and from seeds at any season. In floral form, there are several varieties of Night Jasmine for use. This plant is attacked by number of insect pests. Important among them are listed in the Table 7.

Table 7: Pests of Night Jasmine (*N. arbortristis*).

Sl.No.	*Common Name*	*Scientific Name*	*Family*	*Order*
1.	Moringa caterpillar	*Eupterote mollifera*	Eupterotidae	Lepidoptera
2.	Aphids	*Aphis* sp.	Aphididae	Hemiptera
3.	Thrips	*Thrips* sp.	Thripidae	Thysanoptera
4.	Termites	*Odontotermes*	Termitidae	Isoptera
5.	White grubs	*Holotrichia consanguinea*	Scaraebaiedae	Coleoptera
6.	Gram pod borer	*Helicoverpa armigera*	Noctuidae	Lepidoptera

Moringa Caterpillar (*Eupterote mollifera*) (Figure 23)

Distribution

India and tropical countries

Marks of Identification

Female moth is stout, dark brown and with pectinate antennae and 55-58 mm wing expanse. Males are smaller than females. They measure 48-52 mm in wing expanse and are light yellowish with plumose antennae. Caterpillars are brownish with 13 pairs of hairy tufts on back and measures about 48-50 mm when full grown. Pupa is obtect type, 18-20 mm long and dark brown, tapering at posterior side and broader at anterior side. Eggs are sulphur yellow but, turn brownish yellow at the time of hatching.

Host Plants

Mulberry (*Morus alba* L.), Night Jasmine (*N. arbortristis)* etc.

Life Cycle

Mated female moth lays hundreds of eggs on Night Jasmine leaves, branches/tender twigs/on petiole of leaves in cluster. The clusters are closely fitted with egg masses.

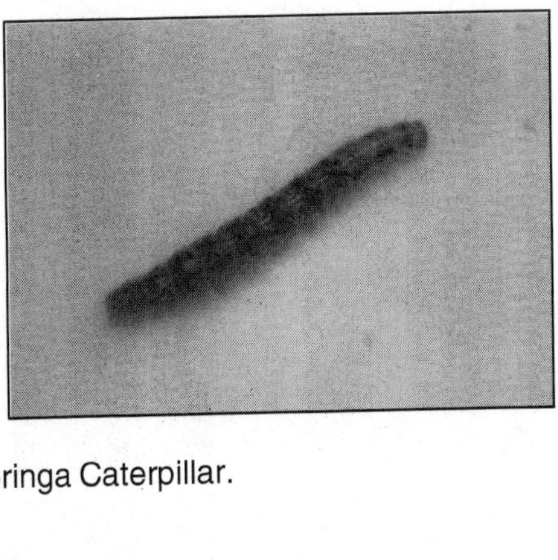

Figure 23: Moringa Caterpillar.

Incubation takes place within 9-13 days. Newly emerged larvae feed on tender leaves of the plant gregariously. The larva moult for 5 times. The larval period is 68 days. Full grown larva pupate in soft, thin cocoon made up of silk and hairs of caterpillars. Pupal period is 35 to 60 days. The life cycle from egg to adult, is completed within 102 to 130 days.

Nature of Damage

The pest become active from August to February during which the pest cause considerable damage to leaves. Caterpillar is the only destructive stage of the pest. Caterpillars feed gregrarious on leaves and skeletonize the plant complete in severe infestation. That affect the flowering bodies adversely.

Control Measures

Preventive

1. Collection and destruction of caterpillars and pupae.
2. Clean cultivation.

Biological

Tachinid fly from Order-Diptera and a Braconid fly, *Dolichogenidea parijatki* Sathe caused mortalities in larval stages of the pest.

Chemical

1. Spray the crop with 0.02 per cent DDVP (or)
   0.02 per cent Rogor (or)
   0.03 per cent Azadirachtin

Aphids, thrips and white grubs are already discussed under previous chapters.

Helicoverpa armigera (Figures 24 and 25)

Larva (Figure 24) feed on leaves.

Adult (Figure 25) is straw coloured with black kidney shaped spot on fore wing.

Figure 24: Larva *H. armigera.*

Figure 25: Adult *H. armigera.*

Chapter 9
PESTS OF GERBERA

Gerbera belongs to the family compositae. It has South African origin. It is famous by the common name, "Transvaal Daisy" or "African Daisy". The best species of Gerbera is *Gerbera jamesonii.*

Gerbera is a dwarf perennial herbaceous plant. The genus *Gerbera* contains about 50 species scattered in Asia and Africa. *G. jamesonii* is very common species widely practiced in India for floriculture. It is stemless, dwarf, 30-45 cm tall and is hairy throughout. The leaves are characterized by long petiole (15 cm), 12-20 cm length, 5-7 cm width and are deeply lobed and located at base as rosette. The flowers are daisy-like, 7-10 cm across with certain exceptions (15 cm across). They are in different colours *viz.,* Lemon yellow, white, cream, orange, pink, salmon, brick red, maroon, scarlet and with different shades. The flower shows a long and slender stalk.

The flowers have long keeping quality in field conditions, in vase and during transportation. Gerbera flowers are very popular for garden decoration, for the vase, for growing mixed borders, rock gardens, etc. Gerbera is also valuable flower for cutting and thus for export and commerce.

Gerbera can be propagated by seeds and by division. It is quite hardy plant and can be grown in plains and in the hills. However, it needs well-drained soil with high moisture but, can be grown in a wide range of soil. Gerbera can be multiplied through tissue culture.

Gerbera is comparatively safe from insects than other ornamental plants. The pests associated with Gerbera are listed in Table 8.

Table 8: Pests of Gerbera

Sl.No.	*Common Name*	*Scientific Name*	*Order*
1.	The mite	*Polyphagotarsonemus latus*	Acarina
2.	The mite	*Steneotarsonemus pallidus*	Acarina
3.	Root knot Nematode	*Meloidogyne incognita*	Nematode

Root Knot Nematode (*Meloidogyne incognita*)

Distribution

Cosmopolitan in distribution. In India, the pest has been reported from South India, Maharashtra, Madhya Pradesh, Uttar Pradesh, Gujarat etc. It is more common in sandy type of soil and under irrigated conditions.

Marks of Identification

The species of *Meloidogyne* genus are commonly called as "Root knot Nematodes". *M. incognita* worms are elongated, about 0.5 to 1 mm in body length. The female adult is milky white, ball or pear shaped. The adult male is also milky white but it is thread like.

Life Cycle

The development of Root knot Nematode consists of three stages, namely egg, larva and adult. The eggs are somewhat oval in shape. They undergo a series of divisions; first stage larvae moult within the eggs and second stage larvae resembles the adults but, they are smaller in size. The second stage feeds on living plants, moult for 3-4 times and become an adult.

Disease Cycle

The second stage female larvae enter into the roots through the holes made by stylet and harbour in subepidermal layer. Immediately, the larva start feeding on the parenchymatous cells of plant. A characteristic knot appears on the roots because of the hypertrophy and hyper plasia induced by nematodes. The larva moult for 4 times in roots only to become matured oval spherical egg laying female. A single female lays about 200-300 eggs. The eggs are elliptical and covered with gelatinous matrix. After hatching from the eggs, the larvae find their body in the soil. The male larvae are parasitic in nature. The males leave the host and enters the sac-like matrices of the female for copulation. The life cycle is completed within 30-40 days. There are 2-3 generations in a single year.

Host Plants

Gerbera, China rose, Mulberry, Potato, Papaya, Jute, Okra, Groundnut, Brinjal, Tomato etc. The nematode infests more than 45 host plants (Sathe, 1998).

Nature of Damage

The nematodes cause the damage to roots by forming characteristic knots/galls. The galls (knots) formed on the roots interfere with the utilization of water and fertilizers effectively. Hence, growth of Gerbera is adversely affected by affecting quality of flower. The affected plants show stunted growth, marginal necrosis and yellowing of leaves and colour change in flowers. The nematode is also responsible for disease complexes with association with fungi, bacteria and viruses, etc.

Control Measures

Preventive

Ploughing or digging the gerbera garden in summer to expose eggs and larvae of the pest to direct sun. The high temperature and low humidity will kill the nematodes.

Cultural

Intercropping with nematocidal plants is useful strategy of pest control.

1. The marigold *Tagetus patula* is grown between Gerbera rows at 10/sq mt when the nematodes enter into the root system of marigold, they get imprisoned because of formation of thick coat of antinematic substances (Belcher & Hussey, 1977).
2. The sunhemp *Crotalaria spectabilis* can also be used as intercropping with gerbera. The same effect is observed with this plant against *M. incognita*.

Chemical

1. Use of organic oil–cakes as soil amendments:

The neem oil cake is effective against root knot nematodes. Apply oil-cake at 1 tonne/ha/year in four equal split doses.

2. Application of nematicides:
   a) Apply–Temik–10 G (Aldicarb) (or)

   Furadon–C (Carbofuran) at 30 kg/ha/year in four equal splits with fertilizer

   b) Apply–Rugby 10-G at 20 kg/ha/year.
3. Mulching of green leaves can suppress the pest population
   a) Mulching of green leaves of neem, *Azadirachta indica* and Pongamia, *Pongamia pinnata* at 1 tonne/ac/crop is effective against Nematode root knot.

Gerbera Mites

1. *Polyphagotarsonemus latus*
2. *Steneotarsonemus pallidus*

Mites cause the damage by sucking the cell sap from tender portions of the Gerbera plant and affect the growth and quality of plant and flowers. The leaves become rusty, flowers get disfigured.

Control Measures

1. Spray the crop with Thiodon 0.05 per cent or Zolone 0.05 per cent or
2. Spray the crop with Dimite (or) chlorobenzilate (or) Kelthane

Aphids & Thrips

These are discussed earlier.

Control

Spray – 0.03 per cent Malathion or 0.02 per cent phosphamidon.

Chapter 10
PESTS OF GLADIOLUS

Gladiolus belongs to the family Iridaceae. It has the origin from South Africa. It is famous by the common names Sword Silly and Corn Flag. Gladiolus is characterized by its attractive spikes having florets of huge form, dazzling colours, varying sizes and long keeping quality. It was first brought to cultivation during Greek period in South Africa. In 18th century Herbert (1806) (England) produced first interspecific hybrid. Later in 1837, Belgium developed Gandavensis hybrids which was found quite useful in development of floriculture and modern gladioli. There are about 10 species of the genus *Gladiolus* which prominently used in Horicultural business at global scenario refer to:

1. *Gladiolus aurantiacus*
2. *G. bladus*
3. *G. cardinalis*
4. *G. tristis*

5. *G. cruentus*
6. *G. trimaculatus*
7. *G. dracocephalus*
8. *G. primulinus*
9. *G. psittacinus*
10. *G. byzantinus*

Gladiolus is best suited bulbous plant in India but not taken as like tuberose. However, it is very famous in Holland and other European countries as it ranks next to Tulip only. There are two types of gladiolus. First group contains large sized flowers and second group is butterfly and miniatures. In Butterfly types spikes are small but, with various coloured flowers. Gladiolus has fantastic several colours from black to white, pink, violet, lilac, greenish, smokey, combinations of 3 colours etc. Following are some important cultivars for use in India and elsewere. Oscar, Snow princes, Cornus, Lily Blue, Melody, High Fashion, Dusk, Friendship, Wild Rose, Bis-Bis, etc. Gladiolus can be grown in plains and upto an attitude of about 2,500 m. The best planting time for this crop is between September and October. June planted produce flowers in August and October and November planted crops can produce in the winter. However, no planting is advisable during January and February because flowers appear in hot months, difficult to maintain.

Propogation of gladiolus is through corms (often wrongly called bulbs) and through seeds. Gladiolus is subject for attack of large number of insect pests. Important pests of gladiolus are recorded in Table 9.

Table 9: Pests of Gladiolus.

Sl.No.	*Common Name*	*Scientific Name*	*Family*	*Order*
1.	The Aphid	*Aphis gossypii*	Aphididae	Hemiptera
2.	Potato aphid (rose)	*Macrosiphum euphorbiae*	Aphididae	Hemiptera
3.	Thrips	*Taeniothrips simplex* *T. traegardhi*	Thripidae Thripidae	Thysanoptera Thysanoptera
4.	Grape mealy bug	*Pseudococcus* sp.	Pseudococcidae	Hemiptera
5.	Tarnished plant bug	*Lygus lineolaris*	Lygaedae	Hemiptera
6.	White grubs	*Holotrichia* spp.	Melolonthidae	Coleoptera
7.	Wire worms	*Elaterid*	Elateridae	Coleoptera
8.	European corn borer	*Ostrinia nubilalis*	Pyralididae	Lepidoptera
9.	Two spotted mite	*Tetranychus urticae*	Tetranychidae	Acarina
10.	Bud mite	*Rhizoglyphus echinopus*	Tetranychidae	Acarina
11.	Root knot Nematode	*Meloidogyne incognita* *M. hapla*	Heteroderidae	Nematoda/ Tylenchida

European Corn Borer (*Ostrinia nubilalis*)

Distribution

Europe, USSR, Japan, Mexico, North America, India etc.

Marks of Identification

Wing expanse of female is 31.75 mm, wings are pale yellow to light brown. Male is smaller than female and with darker wings and slender body. Pupa is obtect type, about 13-19 mm long, brownish, tapering posteriorly and broader at anterior side. Larva when full grown, measures about 25.4 mm, flesh and cream coloured and with faintly spotted dorsum. However, newly emerged larva is pale yellow with black head and measures about 1 mm in body length. Eggs are flattened, scale like.

Host Plants

Gladiolus, corn, jowar, potato, bean, beet, celery hemp, hemp, soyabean, aster, dahalia etc. The pest infects over 200 host plants in the world.

Life Cycle

Female moth lay eggs on tender leaves of *Gladiolus*. Incubation period is 3-12 days. A single female can lay about 400-600 eggs on the under side of gladiolus leaves. Newly emerged larvae wander and feed on tender surface of leaves. Later, they feed on sheath and midrib of leaves, bore into midrib and central shoot and further main stem of the plant. It kills the central shoot, resulting in no sprouting of flowering bodies. The larva moult for 5 times and pupate in the tunnel of plant. Larval period is 25 to 45 days. Pupal period is 8 to 11 days. The pest may complete 1 or 2 generations on gladiolus and then migrate to maize.

Nature of Damage

The larvae feed by boring into midribs, central shoot and main stem, as result no sprouting of flowering bodies are seen. The pest cause 'dead hearts' *i.e.* killing of central shoot.

Control Measures

Preventive

1. Collection and destruction of dead hearts.
2. Collection and destruction of entire plant along with larvae/pupae.
3. Stubbles or left over part after harvest be removed and destroyed/disposed.

Chemical

1. Spray the crop with carbaryl or Fenitrothion.
2. Spray the crop with mixture of sevin and kelthane weekly. This will control pests and diseases of gladiolus.

Aphids, Thrips

Mealy bugs and white grub are already discussed in previous chapters. Follow the control measures as suggested in previous chapters.

Mites and Nematodes

These are covered under the pests of Gerbera.

Wire Worms: Elaterid

Distribution

Widely distributed in Europe. Rarely in Asia and India.

Marks of Identification

Beetle Elaterids show–hind angles of thorax usually produced backwards. A prosternal process is received in the mesosternum. Antennae are often serrate or pectinate.

Grubs or larvae are wire like hence the name wire worms. Grubs are cylindrical, elongated and reddish brown. The segments of their bodies are smooth and fitted closely to one another. The head and rest of the body forms a smooth flexible cylinder. There are 3 pairs of legs and hind end terminates in hooks and chintinized processes which help larva in collecting food material and facilitate movement in the soil.

Nature of Damage

Wireworms feed on corm and roots of gladiolus. They also bore holes in the neck or the base of the leaves.

Control Measures

1. Ploughing and digging the field to expose wireworms to natural mortality factors.
2. Treating soil with 10 per cent BHC dusting.
3. Spray 0.05 per cent phosphamidon/diazinon.

Tarnished Plant Bug

1. *Lygus lineolaris*
2. *Lygus* sp.

Distribution

USA, Europe, India.

Marks of Identification

Adults are mottled, yellowish or reddish brownish, flat or oval, ¼ inch in length and half as wide. Nymphs are greenish in body colour.

Life Cycle

Eggs are laid on leaves. After hatching the eggs, newly emerged nymphs suck the cell sap from leaves and flowers and moult for 4 times and become the adults.

Host Plants

Gladiolus, Marigold and other ornamental plants, certain weeds.

Nature of Damage

Both nymphs and adults suck the cell sap from flower buds causing them to blast or to develop abnormally and to make them unsalable. Growth of the plant is also adversely affected.

Control Measures

Preventive

1. Collection and destruction of eggs, nymphs and adults.
2. Destruction of weeds where pest can breed.

Chemical

1. Spray the crop with 0.1 per cent cygon or sevin just as flower buds begin to develop (or)
2. Spray the crop with 0.05 per cent Rogor (or)
3. Spray the crop with 0.03 per cent Malathion.

Gladiolus Thrips (*Taeniothrips simplex* Morison)

Distribution

India, Tasmania, Many parts of the world.

Marks of Identification

Adults are dark brown or black, ½ inch long, narrow

with fringed wings. Nymphs are pale yellow, smaller than adults.

Life Cycle

Kidney shaped eggs are embedded in plant tissue. Incubation period is 5-6 days, nymphal period is 8 days and pupal period is 18 days.

Host Plant

Gladiolus.

Damage

Nymph and adult suck cell sap from flowers, white dirty speaks are developed on flowers.

Control Measures

Spray 0.05 per cent Malathion.

Chapter 11
PESTS OF DAHLIA

Dahlia belongs to the family Compositae. It has the origin from Mexico. Dahlia is about 30 to 180 cm tall depending upon the cultivar. Its flower consists a certain number of outer ray florets in which male organs are modified into a strap-shaped petal, arranged round a central disc of bisexual florets. Modern era of dahlia started with cross breeds amongst following six species:

1. *Dahlia crocea*
2. *D. variabilis*
3. *D. imperialis*
4. *D. merckii*
5. *D. juarezii*
6. *D. coccinea* (syn. *D. cervantesii*)

Modern Dahlia are classified into 11 types, namely:

1. Single flowered
2. Star
3. Anemone-flowered
4. Decorative
5. Collarette
6. Paeony-flowered
7. Cactus
8. Pompon
9. Double show and fancy
10. Dwarf bedding and
11. Miscellaneous

Dahlia is propagated by seeds, vegetative method by the division of tuber; cutting etc. Dahlia is quite popular as garden flower and compared with Rose for its popularity. It is very good potted plant. It has great potential as cutflower, dahlia beds, mixed borders, for planting in front of the shrub beries or between newly planted shrubbery to fill up the interspaces and bring more colour.

Dahlia is infected by good number of insect pests. They are summarized in Table 10.

Aphids, European Corn Borer

Thrips, mites and nematodes are discussed in previous Chapter.

Leaf Hopper (*Empoasca fabae*)

Distribution

India, Mexico etc.

Table 10: Pests of Dahlia

Sl.No.	*Common Name*	*Scientific Name*	*Family*	*Order*
1.	Aphids	Lean aphid, green peach, leaf curl aphid	Aphididae	Hemiptera
2.	European corn borer	*Ostrinia nubilalis*	Pyralididae	Lepidoptera
3.	Stalk borer	*Papaipema nebris*	Pyralididae	Lepidoptera
4.	Bur dock borer	*P. cataphracta*	Pyralididae	Lepidoptera
5.	Bugs	*Lygus lineolaris* *Poecilocapsus lineatus*	Lygaeidae	Hemiptera Hemiptera
6.	Thrips	*Thrips tabaci* *Heliothrips haemorrhoidalis* *Frankliniala tritici*	Thripidae Thripidae Thripidae	Thysanoptera Thysanoptera Thysanoptera
7.	Leaf hopper	*Empoasca fabae*	Cicadellidae	Hemiptera
8.	Mulberry white fly	*Tetraleurodes mori*	Alerodidae	Hemiptera
9.	Two spotted mite	*Tetranychus urticae*	Tetranychidae	Acarina
10.	Cyclamen mite	*Steneotarsonemus pallidus*	Tetranychidae	Acarina
11.	Root knot Nematodes	*Meloidogyne incognita*	Heteroderidae	Nematoda
12	Snail & Slugs	*Helix* spp. *Achatina fulica* *Laevicantis alti*		Mollusca Mollusca Mollusca

Marks of Identification

Jassids are 2 mm long, pale greenish 1/8 inch long, wedge shaped, walk diagonaly, show piercing and sucking type of mouth parts. Nymphs are similar to adults except lacking wings and they are miniatures. Eggs are bean shaped.

Life Cycle

Eggs are laid in plant tissue, Nymph after emergence from the egg starts sucking the cell sap and moult for 3 times and become adult.

Nature of Damage

Nymphs and adults suck the cell sap from tender leaves and flowers affecting the shape and colour of the leaves and flowers.

Host Plants

Cotton, Dahlia, Vegetables etc.

Control Measures

1. Spray the crop with 0.03 per cent Rogor, (or)
   0.03 per cent Malathion (or)
   0.03 per cent Azadirachtin (or)
   0.02 per cent DDVP (or)
   0.02 per cent Phosphamidon.

Mulberry White Fly (*Tetraleurodes mori*)

Distribution

USA, Canada, India.

Marks of Identification

Adults are small insects looking like small moth, white, and active. Nymphs are elliptical, jet black edged, with whitish fringe of waxy filament. The nymph measures about 1/35 inch in body length.

Host Plants

Mulberry, Guava, Dahlia etc.

Life Cycle

Eggs are laid on Dahlia leaves/flowers. After hatching the egg, newly emerged nymph suck the cell sap from leaves and flowers and moult for 4 times to become pupa and finally adult. Many generations are possible in a season/ year.

Nature of Damage

Both nymphs and adults suck the cell sap from leaves and flowers and disfigure the flowers.

Control Measures

Preventive

1. Infested plant parts should be collected and destroyed along with pest stages.

Chemical

1. Spray the crop with 0.03 per cent Malathion (or)
2. Spray the crop with 0.02 per cent Phosphamidon (or)
3. Spray the crop with 0.03 per cent Resmethrin.

Dialeurodes dispersus

Shows spiral white filaments on the nymphal forms. Adults are white, small moth like active at evening and morning.

Control Measures

As above.

Stalk Borer

1. *Papaipema nebris*
2. *P. cataphracta*

Distribution

Mexico, USA, Canada, India.

Marks of Identification

Moths are small sized, larvae bore the stem.

Host Plants

Dahlia, weeds etc.

Life Cycle

Eggs are laid on the Dahlia plants. After hatching the eggs, newly emerged larvae bore into the shoots/stalk and affect quality and quantity of flowers. The larva moult for 5 times and pupate in stalk only. Pupal period is 6-10 days. From pupa adult moth emerges.

Nature of Damage

Larva bore stalk, kill central shoot/affect quality and quantity of flowers.

Control Measures

Preventive

1. Collection and destruction of bored parts along with larvae and pupae.
2. Removal of harvested crop residues.

Chemical

Spray the crop with the mixture of methoxyclor + Kelthane (1:1) in late June, repeat the dose as per need.

Potato Rot Nematode (*Ditylenchus destructor*)

Distribution

U.S.A., India.

It attacks roots of Dahlia. The infected dahlia shows following features

1. All tuber roots having a cortex with unusual transverse and longitudinal cracking.
2. Sloughing or flaking of the surface or rotted areas should be examined for the presence of this pest.

Control Measures

Hot water treatment to tubers is effective.

Chapter 12
PESTS OF CARNATION

Dianthus caryophyllus belongs to the family Caryophyllaceae. Carnation is native of southern France. It is annual or biannual flowering plant. Dianthus means divine flower. *Dianthus caryophyllus* was in cultivation since 300 B.C. Carnations are of different kinds grown in cooler parts of plain and at medium to high elevation. There are 3 kinds of carnations.

1. *Border carnations and picotees*: Dwarf and branched at base, petals sub divided, picotees–narrow or broad rounded and with smooth petals.
2. *Perpectual flowering*: American tree carnations; bear flowers throughout year, many hybrids are available. Flowers are self coloured, stems longer and branched. Petals are serrated or fringed.
3. *Magnerite carnations*: Represent clove scented flowers, all shades of colour present, petals are fringed. Chabaud carnations are famous.

Good seeds of carnations are difficult in India. Chabaud cultivars include "Aurora", Jeanne Dionis (white), King cup (Yellow), Crimson model (Crimsen) while perpetual flowering cultivars are "Ancient Rose", "Canadian pink", "All wood's crimson", etc. carnations can be propogated sexually, by cuttings and layers. Carnations can be grown in the hills as perennials. The plant can lasts for 3-4 years. However, 4th year carnations are discarded many times. After sowing flowering is possible within 4-6 months only. Carnation has commercial value hence widely taken in India. Pests of carnations are listed in Table 11.

Cut Worms (*Agrotis ypsilon*)

Distribution

World wide: America, Europe, North Africa, Australia, Asia: Japan, China, India, Myanmar, Sri Lanka.

Marks of Identification

Adult body length is about 25 mm. The moth is with dark or blackish colour with some grayish patches on the back. Fore wings contain dark streaks. Full grown caterpillar measures about 44 mm in body length. Plumpy and greasy caterpillar is dark or dark brown in body colouration and with black spots on the back. Young caterpillar is 1.5 mm long and light coloured.

Host Plants

Polyphagous: Carnations, Dahlia, Lilies, Lawns, Gram, Peas, Wheat, Lentil, Linseed, Maize, Jowar, Bhang, Tobacco, Potato, Vegetables, Pulses and several weeds.

Life Cycle

A single mated female can lay about 200 to 350 eggs. Eggs are laid either in the soil or under surface of leaves of

Table 11: Pests of Carnations.

Sl.No.	Common Name	Scientific Name	Family	Order
1.	Cut worms	*Agrotis ypsilon*	Noctuidae	Lepidoptera
		Peridroma saucia		Lepidoptera
2.	Green peach aphid	*Myzus persicae*	Aphididae	Hemiptera
3.	Cabbage Looper	*Trichoplusia* sp.	Noctuidae	Lepidoptera
4.	Other caterpillars	Oblique banded leaf roller–	Pyraustidae	Lepidoptera
		Green house leaf tier	Pyraustidae	Lepidoptera
5.	Onion thrips	*Thrips tabaci*	Thripidae	Thysanoptera
6.	Two spotted mite	*Tetranychus urticae*	Tetranychidae	Acarina
7.	Root knot Nematode	*Meloidogyne incognita*	Heteroderidae	Nematoda

Carnations. Incubation period is 2-12 days depending on climatic conditions. In colder weather incubation period is delayed. Larval period is 25-35 days depending upon climates. Pupal period varies from 10 days (summer) to 30 days (winter). Thus, life cycle of this pest is completed within 45 to 75 days. However, pest is disappeared from April to August.

Nature of Damage

Caterpillar cut down the seedlings of carnation and feed upon them at night.

Control Measure

Preventive

1. Ploughing and digging the field for exposing eggs, larvae and pupae to natural mortality factors.
2. Heavy irrigation or flooding kill caterpillars.
3. Collection & destruction of pest stages *i.e.* eggs, larvae, pupae and adults.
4. Heaps of grasses be kept in the garden as shelter for caterpillars and further, they may collected and destroyed.

Chemical

1. Spray the crop with 0.1 per cent Carbaryl (or)

   0.03 per cent Malathion (or)

   0.03 per cent Azadirachtin

   0.03 per cent DDVP
2. Application of Aldrin or heptachlor dusts @ 2.5 kg/ha.
3. Apply Diazinon to soil prior to planting of crop.
4. Spray the crop with Sevin.

Mechanical

Light traps are useful.

Green Peach Aphid (*Myzus persicae*) (Figure 26)

Distribution

India, USA, Mexico etc.

Marks of Identification

Aphids are soft bodied, louse like insects. They show two pairs of transparent wings, Fore wings are approximately twice the length of hind wing. Nymphs are smaller in size and without wings. Both winged and non winged adults are seen. In both cases, on nymphs and adults a pair of bars or cornicles is present on the abdomen. Adults are 2–2.5 mm long, yellowish green.

Host Plants

Carnations, Cruciferous plants, bean, peach, potato, mustard etc.

Figure 26: Aphids.

Life Cycle

Winter is passed in black shining eggs on host plants. Young aphids seen in spring. They are full grown in about 5-10 days and give birth to wingless forms. Aphids can reproduce vivipariously and parthenogenetically. Many generations are completed during a year.

Nature of Damage

Both nymphs and adults suck the cell sap from leaves and flowers of carnations and affect the quality of flowers adversely.

Control Measures

Preventive

Collection and destruction of infected carnations along with aphids.

Chemical Control

1. Spray the crop with 0.08 per cent Malathion (or)
   0.03 per cent Rogor (or)
   0.02 per cent Diazinon

Onion Thrips (*Thrips tabaci*)

Distribution

World wide distribution.

Marks of Identification

Adults are yellowish-brown, 1 mm long, slender bodied, wings narrow and fringed. Nymphs are light coloured, miniatures and without wings.

Host Plants

Carnations, marigold, onion, garlic, melon, turnip, cucumber, cabbage, cauliflower, bean etc.

Life Cycle

Kidney shaped eggs are embedded into the plant tissue. Incubation period is 5-10 days. A single mated female can lay about 60 eggs. There are 4 instars. Larval period is 4-6 days, pupal period is 2-4 days. Life cycle is completed within 12 days. Many generations are possible in a single year.

Nature of Damage

Adults and nymphs scrape their mouth parts on petals and tender leaves of carnation resulting disfiguring and discolouring of the flowers and leaves.

Control Measures

1. Spray the crop with 0.03 per cent Malathion or sevin

   0.02 per cent phosphamidon (or)

   0.03 per cent Monocrotophos.

Other Caterpillars

Other caterpillars can be controlled by spraying:

1. 0.03 per cent Malathion (or) 0.03 per cent Azadirachtin.
2. Collection and destruction of caterpillars by hand picking.

Nematodes & Mites

Details given in previous chapter.

Chapter 13
PESTS OF CHINA ASTER

China Aster *Callistephus chinensis* belongs to the family Compositae (=Asteraceae). It has origin from China and Japan. Its common name is Aster. It is introduced in floriculture in early 18th century from China. Originally the plant was single flowering, branching type and 60 cm tall. After introduction of this plant in Europe, it has been gone in tremendous changes and developments, as 15 cm–1 m height and pompon flowers about the size of a button to large flower heads having single, double, anemone-flowered, peony flowered, incurred, quilled or shaggy flower types etc. It is one of the most valuable garden plant because of variation in flower colours. The colours include pure white, pink shades, pale blue, dark blue, scarlet, prim rose, mauve, purple etc. However, pure yellow coloured flowers are not produced by this plant. The flowers last long in water and have bedding and potting importance. Following types of China Asters are available for floriculture.

1. Manmoth Peony flowered
2. Victoria
3. Pompon
4. Comet
5. Giant Californian
6. Chrysanthemum
7. Branching aster

China aster is propogated by seeds. For continuous flowering sowing should be done at 7-10 days interval. It is quite important as cut flower treads.

China Aster is attacked by several pests. Those are listed in Table 12.

Leaf Hopper (*Macrosteles fascifrons*)

Distribution

U.S.A., Mexico, India.

Marks of Identification

Jassids are wedge shaped, about 2–2.5 mm long, walk diagonally. Nymphs are miniatures and lacking wings.

Host Plants

China Aster, Cotton, Malvaceous plants.

Life Cycle

Bean shaped eggs are embedded into the soft plant tissues. After hatching the eggs, newly emerged nymphs start sucking the cell sap from tender portions (buds, leaves, flowers) and moult for 3 times to become adults.

Nature of Damage

Adults as well as nymphs suck the cell sap from plants

Table 12: Pests of China Aster.

Sl.No.	*Common Name*	*Scientific Name*	*Family*	*Order*
1.	Leaf hopper	*Macrosteles fascifrons*	Jassidae	Hemiptera
2.	Corn root aphid	*Aphis maidiradicis*	Aphididae	Hemiptera
3.	Potato aphid	*Macrosiphum solanifolii*	Aphididae	Hemiptera
4.	Asiatic garden beetle	*Maladera castanea*	–	Coleoptera
5.	Black blister beetle	*Epicauta pennsylvanica*	Meloidae	Coleoptera
6.	Japanese beetle	*Popillia japonica*	Melolonthidae	Coleoptera
7.	Tarnished plant bug	*Lygus lineolaris*	Lygaeidae	Hemiptera
8.	Two spotted mites	*Tetranychus urticae*	Tetranydridae	Acarina
9.	Root knot nematodes	*Meloidogyne incognita*	Heteroderidae	Nematoda
10.	Leaf rollers, Mealy bugs, beetles		–	

as a result, leaves and petals becomes curly and disfigured and discoloured which ultimately affect the quality and marketability of carnations.

Control Measures

1. Spray the crop with 0.03 per cent Malathion or
   0.02 per cent Diazinon or
   0.02 per cent Metasystox or sevin.

Corn Root Aphid (*Aphis maidiradicis*)

Distribution

Mexico, USA, India.

Marks of Identification

Adults are pale grayish, nymphs lighter in colour. Aphids are soft bodied, pear shaped, with a pair of cornicles on the abdomen and cauda at the tip of abdomen. Nymphs are miniatures and lacking wings.

Host Plants

Carnations, Maize, Jowar, graminaceous weeds etc.

Life Cycle

Aphids reproduce parthenogenetically females. They can also reproduce viviparously, ovovivipariously and compete many generations in a single year.

Nature of Damage

Nymphs as well as adults suck the cell sap from roots of the plant. Inject toxins into the plant body resulting in curly leaves and petals of the flowers. The pest also secrete honey dew like substance which create sooty mould on the leaves and flowers and affect the quality of flowers.

Yellowing, dwarfing and wilting of the plant are some very important aspects of pest damage. Root knots are formed to plant roots by the aphids.

Control Measures

Preventive

Collection and destruction of infested crop parts along with pest stages.

Chemical

1. Spray the crop with 0.03 per cent Malathion (or) Azadirachtin or

   0.02 per cent Phosphamidon (or)

   0.03 per cent Rogor

Potato Aphid (*Macrosiphum solanifolii*) (Figure 26)

Distribution

India, Mexico, USA

Marks of Identification

Louse like aphids are soft bodied, winged or non winged and with a pair of siphunculi. Nymphs don't show wings. They are smaller than adults.

Life Cycle

Reproduce parthenogenetically and viviparously. Many generations are completed in a year.

Nature of Damage

Both nymphs and adults suck the cell sap from leaves, stems and blossoms of the china asters affecting the quality of flowers adversely.

Control Measures

Preventive

As above species.

Chemical

As above species.

1. Spray 0.03 per cent Metasystox or sevin (or) 0.03 per cent Malathion.

Natural

Lady bird beetles feed on aphids.

Asiatic Garden Beetle (*Maladera castanea*)

Distribution

Asia: India.

Marks of Identification

Beetles are brown in colour. Grubs are whitish, pupa is exarate type.

Host Plants

Carnations and other ornamental plants in the garden.

Life Cycle

Eggs are deposited in soil. Newly emerged grubs feed on roots of the crop and various other plants and also on humus material. The full grown grub pupate in soil. Finally, the adult formed can stay in the soil and come out of soil at night only. Only one generation is possible in a single year.

Nature of Damage

Beetles feed on foliage at night. During day time they buried themselves into the soil near the bases of the crop plant. Damage flowers at night.

Control Measures

Preventive

1. Ploughing and digging the field for exposing immature stages to natural mortality factors (biotic and abiotic).
2. Collection and destruction of beetles.
3. Beetles get attracted to light. Therefore, use light traps for catching and killing them.
4. Apply diazinon to soil for killing grubs and pupae.
5. Spray the crop with sevin as per the need or incidence.

Black Blister Beetle

1. *Epicauta* sp.
2. *Lytta* sp.

Distribution

Mexico, USA, India.

Marks of Identification

Adult beetles are ½ inch long and black, metallic shining. They form blister on human skin when crushed on the body. Grubs are whitish, pupae are brownish and exarate type.

Host Plants

Carnations, chrysanthemum, clematis, zinnia etc.

Life Cycle

Eggs are laid in soil. Grubs develop in soil. Pest pupate in soil.

Damage

Beetles feed on flowers.

Control Measures

Preventive

As above species.

Chemical

Spray methoxychlor or sevin.

Japanese Beetle

Already given in previous chapter.

Tarnished Bug

Mites and nematodes are already discussed in previous chapters.

Chapter 14
PESTS OF ORCHIDS

Orchids belongs to the family Orchidaceae. They are flowering plants. Although Chinese are known to grow orchids but, the first orchid, *Bletia purpurea* was grown by Englishman around 1731. There are about 24,000 species of orchids and 32,000 hybrid forms. Hybrids are released in day to day market at the rate of 1,000 per annum. Thus, approximately 3 hybrids are released per day in the floriculture market.

Orchids are perennial herbs. They may be epiphytes (air-plants), lithophytes (growing or rocks) terrestrial (ground dwelling) or saprophytes (growing on decaying organic mater). *Galeola falconeri* is Indian largest orchid with 40-45 cm rhizomes of 2.2–2.5 m height while, *Eria pusilla* is the smallest orchid measuring about 1 cm as a whole plant. Orchids are found in all parts of the world including snow covered peaks of Himalayas.

Orchids have highly decorative flowers which lasts for few days to a few months. Many commercial hybrids are produced from the genera *Vanda, Cymibidium, Cattleya, Dendrobium, Paphiopedium* etc. Sale of flowers and plants is multimillion dollar business in USA and other countries like Thailand, Singapore, Malaysia and Philippines. Indian orchid floriculture is not satisfactory. Indian do not multiply orchids but just collect from natural habitat. This has resulted severe damage to a natural wealth of our country and no improvement in new hybrids. Because, very economic species such as *Dendrobium densiflorum, Thunia alba, Coelogume cristata, Symbidium elegans, Vanda coerulea* etc. are almost at the verge of being extinct.

Orchids have several uses in day to day life. The medicine salep is obtained from some orchids (Genus *Orchis*). In China, the drug, "chin shihhu" is obtained from dried stems of *Dendrobium nobile*. It is antipuretic and tonic. Large number of alkaloids also been isolated from various orchids. "Dendrobine" is common alkaloids derived from many orchids. In Malaysia orchids are used as vegetables. Beverage and tea is also prepared from some orchids particularly dried and curved leaves and pseudobulbs are used in this regard. Weaving baskets are prepared from orchids in Indonesia. Similarly, musical instruments like guitar is prepared by using some orchids (*Cyrtopodium, Geodorum*) in Philippines and Brazil.

Insect pests affect the orchid growth and flowers adversely. Therefore, their control is essential part of good floriculture. Important pests of orchids are listed in Table 13.

Table 13: Pests of Orchids.

Sl.No.	*Common Name*	*Scientific Name*	*Family*	*Order*
1.	Aphids	*Cerataphis orchidearum*	Aphididae	Hemiptera
		Neomyzus circumflexus	Aphididae	Hemiptera
2.	Orchid weevil	*Diorymerellus laevimaryo*	Curculionidae	Coleoptera
3.	Orchid bulb borer	*Metamasius graphipterus*		Lepidoptera
4.	Orchid fly	*Eurytoma orchidearum*	Eurytomidae	Hymenoptera
5.	Orchid mealy bug	*Pseudococcus microcirculans*	Pseudococcidae	Hemiptera
6	Cattleya midge	*Parrallelodiplosis catleyae*	Cecidomyidae	Diptera
7.	Cattleya weevil	*Cholus cattelyae*	Curculionidae	Coleoptera
8.	Jendrobium borer	*Xyleborus morigerus*		Coleoptera
9.	Orchid plant bug	*Tenthecoris bicolor*		Hemiptera
10.	Thrips	*Chaetanaphothrips orchidii*	Thripidae	Thysanoptera
		Heliothrips haemorrhoidalis	Thripidae	Thysanoptera
		Hercinothrips femoralis	Thripidae	Thysanoptera
11.	Scales	*Scales*	Coccidae	Hemiptera
12.	Slugs, Nematodes, mites	–	–	Non insects

Orchid Aphid [*Cerataphis orchidearum* (Westwood)]

Distribution

India, Tropics, Europe and North America

Marks of Identification

Adults of nonwinged forms are sedentary, small, flattened, broadly oval, dorsal surface dark reddish brown to black dusted with wax, about 1.5 mm long, while winged forms are about 2 mm long.

Host Plants

Orchids.

Life Cycle

These species are anholocyclic. Sexupare are not known, reproduce parthenogenetically. Many generations are completed in a single year.

Nature of Damage

Nymphs and adults suck the cell sap from orchids and affect the beauty of orchids by disfiguring and discolouring. The pest also secret honeydew like substance which invite sooty mould to form on the plant, that affect photosynthesis and growth of the plant adversely.

Control Measures

As suggested for aphids in previous chapter.

Orchid Mealy Bugs (*Pseudococcus microcirculans*)

Distribution

Tropics, Europe, USA (California, Florida), India.

Marks of Identification

Mealy bugs are flat bodied, orange or mahogany

coloured insects which are covered with cottony, mealy like powder on the body. Males are winged and females and nymphs are non-winged.

Hosts

Ornamental plants, orchids.

Life Cycle

The pest reproduce viviparously and partheno-genetically.

Nature of Damage

Both nymphs and adults suck the cell sap from the cactus and affect the quality of flowers and entire plant.

Control Measures

Preventive

1. Collection and destruction of infected plant parts along with pest stages.

Chemical

1. Spray the crop with 0.03 per cent Malathion/ Rogor.

   0.02 per cent Phosphamidon/dialdrin/diazinon (or)

   0.03 per cent metasystox–R or Cygon.

Biological

Lady bird beetles feed on mealy bugs.

Orchid Plant Bug (*Tenthecoris bicolor*)

Distribution

U.S.A., Europe, Asia.

Marks of Identification

Adult bugs have different colouration from orange to bright red, with a black design down the back. Nymphs lacking wings and are light orange to red colouration.

Host Plants

Orchids.

Life Cycle

Eggs are laid on the plants in clusters. Newly emerged nymphs suck the cell sap from orchids and moult for 4-5 times to become the adults.

Nature of Damage

Both nymphs and adults suck the cell sap from orchids and affect growth of plant and quality of the orchids.

Control Measures

Preventive

Collection of egg masses, nymphs and adults and destruction of them.

Chemical

1. Spray the crop with 0.04 per cent Methoxychlor
   or
   0.04 per cent Monocrotophos

Orchid Weevil

1. *Diorymerellus laevimaryo*
2. *Diorymerellus* sp.

Distribution

U.S.A., India.

Marks of Identification

Adults have snout which is extension of head. Adult weevil is about 1/8 inch long, black, smooth and shining. The grub is typically quama shaped, whitish, 1/16 inch long and legless. Pupa is brownish and exarate type.

Host Plants

Orchids.

Life Cycle

Eggs are laid in soil. After hatching the eggs, newly emerged grubs feed on roots of orchids. Full grown grubs pupate in soil and become an adult by pending one week or two.

Nature of Damage

Weevils can also feed on roots. However, their main damage is to the young and tender leaves, to the sheath surrounding the flower buds and to bulbs. They also feed on flower petals before the flowers open. They prepare numerous holes to the flowering bodies and invites pathogenic attack to plant, resulting in decay in blossoms. Grubs feed on new roots hollowing out the inside and causing the tips to turn black.

Control Measures

Preventive

1. Ploughing or digging the field for exposing life stages of the pests to natural mortality factors.
2. Collection and destruction of weevils.

Chemical

1. Spray the crop with 0.04 per cent Methoxychlor

Cattleya weevil (*Cholus catleyae*)

Distribution

U.S.A.

Marks of Identification

Adult is less than ½ inch long with snout which is extension of head. It has white marks on the back. Grub is whitish, legless and pupa is exarate type.

Host Plants

Orchids.

Life Cycle

Eggs are laid either on plant or in soil. Grub feed on leaves and develop on stems and pseudobulbs. Full grown grub pupates on the plant or in soil and become adult.

Nature of Damage

Weevil adult cause damage to pseudobulbs by feeding on the surface and also puncturing the leaves by feeding. Grubs feed on leaves, stems and pseudobulbs resulting decaying of flowers and/or failure of plant to bear flowers.

Control Measures

As above.

Cattleya Midge

1. *Parrallelodiplosis cattleyae*
2. *Parrallelodiplosis* sp.

Distribution

USA, India.

Marks of Identification

Adult is slender bodied dipterous fly with bed like antennae. Blackish or brownish or orange in body colour. Maggots are legless yellowish and tapering posteriorly. Pupa is barrel shaped.

Host Plants

Orchids.

Life Cycle

Elongated eggs are laid on the plant. After hatching, newly emerged maggot feed on the tips of the roots of orchids. Moult for 3 times for pupation stage and finally moult into the adult.

Nature of Damage

Maggots feed on the tip of the roots and arrest the growth of the plant and also form unsightly nut like galls which affect the beauty of the orchid and food transportation.

Control Measures

Chemical

Spray the crop with 0.04 per cent Methoxychlor.

Mechanical

Midges are attracted to light hence light sticky traps can be used for collection and destruction of the pest.

Orchid Fly (*Eurytoma orchidearum*)

Distribution

USA, India, Europe etc.

Marks of Identification

Adult is small black wasp, 1/8 inch long. It has petiolate abdomen and reduced veination in the wings. Grubs are whitish. Typically tapering towards both the ends. Eggs are elongated and whitish.

Host Plants

Orchids.

Life Cycle

Female deposit its eggs at the bases of the pseudobulbs or some times on young leaves and rhizomes. Incubation period is 1 or 2 weeks. Newly emerged grubs feed on either leaves or pseudobulbs or rhizomes and make tunnels or galleries. The full grown grub pupate inside the galleries and then adults find their way out. During this period the buds turn brown or black and no flowers are formed.

Nature of Damage

As above.

Control Measures

Spray the crop with 0.04 per cent methoxychlor.

Orchid Bulb Borer (*Metamasius graphipterus*)

Distribution

U.S.A.

Marks of Identification

Adults are blackish with large yellowish blotches on the wing covers.

Host Plants

Orchids.

Life Cycle

Female lay eggs on the plant. Larvae feed on leaves and moult for 5 times and pupate. The pupa is transformed into the adult.

Nature of Damage

Larvae feed on leaves and other parts of the orchids. They feed inside the bulb and provide the way to secondary infections to fungi to rot plant part (bulb).

Control Measures

Preventive

Infected parts should be cut down and disposed along with pest stages.

Chemical

Spray the crop with 0.04 per cent methoxychlor.

Scales

Orchid is attacked by several species of scales. Important scales of orchids are given below:

1. Hemispherical scales
2. Black
3. Florida Red
4. Latania
5. Oleander
6. Oyster shell
7. Purple
8. Tea
9. Chaft
10. Cyanophyllum etc.

All above scales suck the cell sap, affect the growth of orchids, affect the beauty of the orchids.

Control Measures

Spray the crop with 0.03 per cent Malathion (or) Sevin. These pesticides are effective at crawling stage (1st instars).

Thrips

Orchids are attacked by varieties of thrips. Thrips rub their mouth parts on flowers or tender leaves/tender parts of the plant and feed on oozing sap resulting in white or dark speaks on flowers, leaves etc. Details of their life cycle is given in earlier chapter. The important thrip species associated with orchids/cattleya refers to:

1. *Heliothrips haemorrhoidalis*
2. *Hercinothrips femoralis*
3. *Chaetanaphothrips orchidi*

Control Measures

Spray the crop with 0.03 per cent Malathion or cygon.

Mite Blotch

Can be controlled by spraying kelthane or Dimite at 2 to 3 week interval before flower appears.

Slugs

Can be controlled by spraying the crop with ½ pint of metaldehyde EC in 10 gallons of water for each 100 square yard/year.

Chapter 15
PESTS OF BOUGAINVILLEA

Bougainvillea (Figure 27) is most popular ornamental shrub widely used in gardens in India. It belongs to the family Nycataginaceae. Bougainvillea was first cultivated by French Botanist. However, it has origin from South America.

Bougainvillea contains simple, alternate, roundish ovate or elliptic–lanceolate leaves. The plant is armed with stout spines in the stem. There are 3 brightly coloured beautiful petal like bracts. In recent days multi bracted races are also available for gardening. The bracts have a variety of colours ranging from white to deep magenta, lighter shades as orange, yellow, purple, pink, scarlet, crimson, red and mauve. In some cases bracts are of two colours in other cases colour of bracts changes with age showing 2-3 coloured plant. However, the actual flowers are small, tubular, ridged and open into a star at the apex. The flower

Figure 27: Bougainvillea .

colouration may be white, light greenish yellow, cream, pink or yellow.

In north India, Bougainvillea has blooming peak during September to December and again from February to June. Like Gulmohar and few others this plant increase the beauty of garden in dry season. Hence, it is believed that the best flowering of this plant is in dry period. Bougainvillea can be grown at higher altitudes upto 1500 to 2000 m as in Nainital, Almora etc. Out of 10 species of Bougainvillea, 3 are commercially utilized in floriculture *viz.,*

1. *B. glabra*
2. *B. spectabilis*
3. *B. peruviana*

It is used in the garden as shrub and as climber. Important cultivars available for use are "Sanderiana",

"R.R. Pal", Thimma, Partha, Mary Palmer, H.C. Buck etc. This plant can be propogated by cuttings, by ground or air layering and budding. It can be grown in different types of soil and climatic conditions.

Pests of Bougainvillea

In India, *Bougainvillea* is less attacked by insect pests. The pests associated with this plant are listed in Table 14.

Thrips (*Thrips* sp.)

Distribution

India.

Marks of Identification

Thrips are slender bodied and with two pairs of wings which are narrow and fringed. Adults are brownish and nymphs are light brown coloured and without wings. Adults are less than 1 mm long.

Host Plants

Bougainvillea and many flowering ornamental plants.

Life Cycle

Kidney shaped eggs are laid in scars prepared in tender buds or bracts or on flowers. After hatching the eggs newly emerged nymphs feed on oozing cell sap of the plant by rubbing their mouth parts on the tender parts of leaves and bracts and flowers. Pupa does not feed. Adults can also cause the damage to crop in the same way as nymphs do.

Nature of Damage

Adults and nymphs cause damage scraping their mouth parts on tender parts and colourful leaves and causing white or brownish streaks or specks on the bracts

Table 14: Pests of Bougainvillea.

Sl.No.	*Common Name*	*Scientific Name*	*Family*	*Order*
1.	Bougainvillea caterpillar	*Asciodes gordialis*		Lepidoptera
2.	Scale insects	i) Brown soft	Coccidae	Hemiptera
		ii) Cottony cushion	Coccidae	Hemiptera
		iii) Florida red	Coccidae	Hemiptera
		iv) Hemispherical	Coccidae	Hemiptera
		v) Latania	Coccidae	Hemiptera
		vi) Mining	Coccidae	Hemiptera
		vii) Pustule	Coccidae	Hemiptera
		viii) Cyanophyllum	Coccidae	Hemiptera
		ix) Qua-hog shaped	Coccidae	Hemiptera
3.	Thrips	*Thrips* sp.	Thripidae	Thysanoptera
4.	Aphid	*Aphis gossypii*	Aphidae	Hemiptera
5.	Bihar hairy caterpillar	*Spilosoma obliqua*	Arctiidae	Lepidoptera

or flowers or leaves and thus affecting the quality and beauty of the plant.

Control Measures

1. Spray the crop with 0.03 per cent Malathion (or) 0.03 per cent Azadirachtin (or) 0.02 per cent Phosphamidon.

Bihar Hairy Caterpillar (*Spilosoma obliqua*) (Figure 28)

Distribution

Oriented region. In India, it is reported from Maharashtra, Karnataka, Kerala, Tamil Nadu, Orissa, West Bengal, Madhya Pradesh, Uttar Pradesh, Andhra Pradesh, Rajasthan, Punjab etc.

Marks of Identification

Adult moths measure about 50 mm in wing expanse. Its head, thorax and under surface is dull yellow. Antennae and eyes are black. Abdomen contain a series of black spots on each segment. On the hind wing about 10-12 black spots are present and arranged in two rows. Caterpillars are

Figure 28: Bihar Hairy Caterpillars.

orange with 2 black patches (posterior and anterior side) and with long grayish hairs. Full grown larva is about 45 mm long.

Host Plants

Polyphagous, many ornamental plants including Bougainvillea, Mahendi, linseed, Mustard, pulses, vegetables, etc.

Life Cycle

A single female can lay about 1000 eggs on plant. Incubation period is 8-13 days. Larva moult for 7 times. Larval period is 4-8 weeks. Pupal period is 1-2 weeks. Life cycle is completed in about 6-12 weeks. The pest can complete 3-4 generations in a single year.

Nature of Damage

Larvae feed gregariously on leaves and bracts for first 2 instars and then feed solitarily. They affect the growth and quality of leaves/bracts and flowers adversely.

Control Measures

Preventive

1. Collection and destruction of egg masses and gregarious and solitary larvae.
2. Clean cultivation and destruction of pupae.

Chemical

1. Spray the crop with

   0.15 per cent Carbaryl (or)

   0.03 per cent Malathion/Azadirachtin

Biological

Release *Meteorus dichomeridis* parasitoids 16,000 ha/ week as per the need.

Scales

Control is given in Chapter 13.

Chapter 16
PESTS OF TUBEROSE

Tuberose *Polianthes tuberosa* (Figure 29) belongs to the family Amaryllidaceae. It has its origin from Mixico. In Hindi it is called as Rajanigandha. Its leaves are light green, grass like, long and narrow. It is 15-20 cm tall. Flowers are pure white and tinged with red in double cultivars (pearl). Variegated cultivar has variegated yellow margins to leaves. The flower is single with red tinge in 2nd stage. Flowering stalk provided from centre of the cluster of leaves, 50-90 cm high. Single tuberose have very high fragrant than double. Flowers can be useful for cutflowers, vase and bouquets, veni, garland, buttonholes, crown, etc. They have great importance in religious, marriage and birthday ceremonies. The spike can stay for 7-12 days in vases. This plant is cultivated as beds, borders or potted plant. Tuberose can be propogated from bulbs and planted during rains.

Table 15: Pests of Tuberose.

Sl.No.	*Common Name*	*Scientific Name*	*Family*	*Order*
1.	Aphid	*Aphis* sp.	Aphididae	Hemoptera
2.	Thrips	*Thrips* sp.	Thripidae	Thysanoptera
3.	Caterpillars	*H. armigera*	Noctuidae	Lepidoptera

Figure 29: Tuberose.

Control of above insects is given in previous chapters. Thrips and Aphids cause damage by sucking the cell sap and disfiguring the flowers. Treating the crop with 0.03 per cent malathion or 0.03 per cent Azadirachtin will control the above pests.

Chapter 17
PESTS OF CACTI

Cactus group of plants belongs to the family cactaceae. All cactus show Areoles or spine-cushions with some exceptions. The fruit is one cell berry, cacti are perennial and flower petals arose from top of the ovary. All cacti belongs to dicotyledones group. Cacti are leaflegs or leaves are vestigial. There is much variation in plant shape, form, size and colour of areoles and thorns. Flowers are metallic shining and very beautiful. The flowers may be white to pink, purple, yellow, crimson, violet, dark crimson red and combinations of various colours. Cacti can be propogated by seeds, cuttings, offsets and grafting. There are several species of cacti and beautiful hybrid varieties of the species.

The cacti are attacked by several insect pests. Important pests of cacti are listed in Table 16.

Table 16: Pests of Cacti.

Sl.No.	*Common Name*	*Scientific Name*	*Family*	*Order*
1.	Mealy bugs	*Pseudococcus longispinus*	Pseudococcidae	Hemiptera
		Planococcus citri	Pseudococcidae	Hemiptera
2.	Ground mealy bug	*Rhizoecus falcifer*	Pseudococcidae	Hemiptera
3.	Aphid	*Dysaphis tulipae*	Aphididae	Hemiptera
4.	Cactus Scale	*Diaspis echinocacti*	Coccidae	Hemiptera
5.	Greedy scale	*Hemiberlesia rapax*	Coccidae	Hemiptera
6.	Lesser snow scale	*Pinnaspis strachani*	Coccidae	Hemiptera
7.	Oleander scale	*Aspidoitus nerii*	Coccidae	Hemiptera
8.	Two spotted mite	*Tetranychus urticae*	-	-

Mealy Bugs

1. *Pseudococcus longispinus*
2. *Planococcus citri*

Distribution

World wide.

Marks of identification, host plants, life cycle, damage and control is given in Chapter 4.

Ground Mealy Bug

Rhizoecus falcifer

Distribution

U.S.A., Mexico

Marks of Identification

Female is flat bodied, wingless, orange, mahogany with white mealy like power on the back. Males are winged.

Life Cycle

Reproduce viviparously, ovoviviparously.

Nature of Damage

Nymphs and only female adults suck the cell sap from terminal roots of cacti. They can also suck the cell sap from roots of potted plants and disfigure the plant and flowers.

Control

As above species.

Aphid [*Dysaphis tulipae* (Bdf)]

Distribution

Almost cosmopolitan. It is not reported from South America.

Marks of Identification

Wingless forms are 1 to 2.30 mm long, very pale yellow, grey and pink. Due to covering of white wax powder they appear as white. Siphunculi are dark. Winged forms measure about 1.5 to 2.3 mm in body length and they are pale yellowish grey or some times pinkish, abdomen contain a central black dorsal patch. Siphunculi are black.

Host Plants

Polyphagous: A very large number of monocotyledonous species of Liliacae, Araceae, Musaceae and Iridaceae and members of the following genera are infected by this species – *Cacti, Lilium, Gladiolus, Crocus, Chionodoxia, Freesia, Arum, Iris,* etc.

Life Cycle

It is entirely anholocyclic. Sexual forms are not found. They spend the dormant period of their host plants under the scales or in crevices of the bulbs, corms or rhizomes. The pest can reproduce parthenogenetically females, many generations are completed in a single year.

Nature of Damage

Both nymphs and adults cause the damage to ornamental and other plants by sucking the cell sap, secreting honeydew like substance, creating sooty moulds and affecting growth of the plant and quality of look of ornamental plants like Cacti, Lilium and Gladiolus etc. It also acts as viral transmitter for viral disease of Lilly and Tulip.

Control Measures

Preventive

1. Collection and destruction of infested plant parts along with pest stages.

Chemical

1. Spray the crop with 0.03 per cent Malathion (or)
   0.03 per cent Rogor (or)
   0.02 per cent Phosphamidon.

Biological

Lady bird beetles and lace wings feed upon aphids in nature.

Cactus Scale (*Diaspis echinocacti*)

Distribution

U.S.A., India.

Marks of Identification

Female scales have grey circular covering while, the males are white and slender bodied and winged individuals.

Host Plants

Cacti.

Life Cycle

Eggs are deposited beneath the female body in egg sac. After hatching the eggs, nymphs come out of the female sheltering body and seek out suitable place for insertion of mouth parts into the plant body. The first instar individuals are crowlers. They settle on a plant and suck the cell sap and moult for 4 times to become an adult.

Nature of Damage

Females and nymphs suck the cell sap from cacti. Males do not cause damage as it does'nt feed. The pest secrete honey dew like sticky substance which create sooty moulds on the plant and affect the growth and quality of flowers, also affect the entire look of the plant adversely.

Control Measures

Preventive

Collection and destruction of infested plant parts along with pest stages.

Chemical

Spray the crop with 0.03 per cent Malathion (or) 0.02 per cent Phosphamidon. For other scales, the control measures are same as recommended for *D. echinocacti.*

Two Spotted Mites (*Tetranychus urticae*)

Distribution

World wide.

Marks of Identification

Cacti infected with mites shows ashy, yellowish or whitish appearance.

Host Plants

Cacti, China Aster, Gladiolus, etc.

Life Cycle

Eggs are laid on plant. Larvae developed into adults on plants by sucking the cell sap.

Control

Spray the crop with chlorobenzilate or kelthane (or) metasystox-R (or) Tedion from under surface of leaves.

Chapter 18
PESTS OF CYCADS

Cycads belong the Family Cycadaceae. Some of plants from cycads are quite beautiful and widely grown in gardens of tropical and subtropical regions of the world. They look like palm but are quite different botanically. Most of the plants are slow in growth. Annual shedding of leaves are seen. There are 9 genera of cycads which are used for gradening.

Cycas, Zamia, Marozamia, Dioon and *Encephalartos* are commonly used in gardens of India. The cycads can be propogated through seeds and suckers. Some insects attack the cycads. The list of which is given in Table 17.

Details are given in Chapter 17.

Table 17: Pests of Cycads.

Sl.No.	*Common Name*	*Scientific Name*	*Family*	*Order*
1	Scales	*Diaspis* sp.	Coccidae	Hemiptera.
2	Mealy bug	*Pseudococcus* sp.	Pseudococcidae	Hemiptera
3	Dracaena thrips	*Heliothrips* sp.	Thripidae	Thysanoptera

Chapter 19
PESTS OF FERN

Ferns belongs to the class Pteridophyta in Plant kingdom. There are about 12,000 species of fern in the world. From India 1,000 species have been reported but only a very small number of species are in cultivation. The plant look like small hair as stem and with fewmoss like leaves. It may be as tall tree. Its height may be above 24 m. The plant has three basic parts *i.e.* root, rizome or stem and leaf (frond). The fern may be terrestrial, epiphytic or aquatic found in ponds, lakes and stagnant pools. The ferns are extremely handsome due to their foliage and hence used in gardening. They are very suitable for shade gardens or may be grown extensively in pots and pans for indoor decoration, preparing bouquets, buttonholes and wreaths. Modern hair ferns are used for pasting on greeting cards. Thus, ferns have unlimited uses and widely cultivated in the world as ornamented plants. The ferns are propogated through spores. Hybrids are also available for use. However, ferns

Table 18: Pests of Fern.

Sl.No.	*Common Name*	*Scientific Name*	*Family*	*Order*
1	Aphids	*Aphis* sp.	Aphididae	Hemiptera
2.	Caterpillars	–	-	Lepidoptera

are attacked by several insect pests. The pests associated with fern plants are listed in Table 18.

The above pests can be controlled by spraying 0.03 per cent malathion/0.05 per cent DDVP.

Chapter 20
PESTS OF MICHELIA

Michelia belongs to the family Magnoliaceae. The genus *Michelia* contains two species *viz.*

1. *Michelia alba* and
2. *Michelia champaca*

Michelia alba has origin from java and is tree variety flowering plant. Flowers are white and faintly scented for a day. *Michelia champaca* has origin from India. Its common name is Golden Champa. In Marathi it is called Sonchampa or Sonchapa. It is medium sized tree of long leaves and golden flowers with high scent and silky appearance. Flowers measures about 5-7 cm in length and leaves are with wary margins and larger than mango leaves. It can bear pods very less in number. It has 10-30 m height. April-May and September to October are the flowering seasons. They can be grown up to medium altitudes and humid atmosphere. The plant comes to flowering stage after

Table 19: Pests of Michelia.

Sl.No.	*Common Name*	*Scientific Name*	*Family*	*Order*
1.	Thrips	Thrips sp.	Thripidae	Thysanoptera
2.	Aphids	*Aphis gossypii*	Aphididae	Hemiptera
3.	Mealy bug	*Pseudococcus* sp.	Pseudococcidae	Hemiptera
4.	Jassids	*Empoasca* sp.	Jassidae	Hemiptera
5.	Scales	Brown scale	Coccidae	Hemiptera

plantation within 7-8 years. It is used in gardens and parks and also grown around house. Flowers are used for making garlard. The plant has medicinal importance. Michelia is propogated by grafting or cuttings.

Thrips (Figures 30–31)

Distribution

India

Marks of Identification

Adult thrips are slender bodied, small insects, measuring about 1 mm in body length. They are blackish brown in body colouration. They contain two pairs of narrow and fringed wings. The tip of abdomen contains sharp and short ovipositor for egg laying in case of female. Nymphs are light brownish in body colouration. They don't show wings and are still smaller than adults but quite active. Eggs are bean shaped.

Figure 30: Thrips.

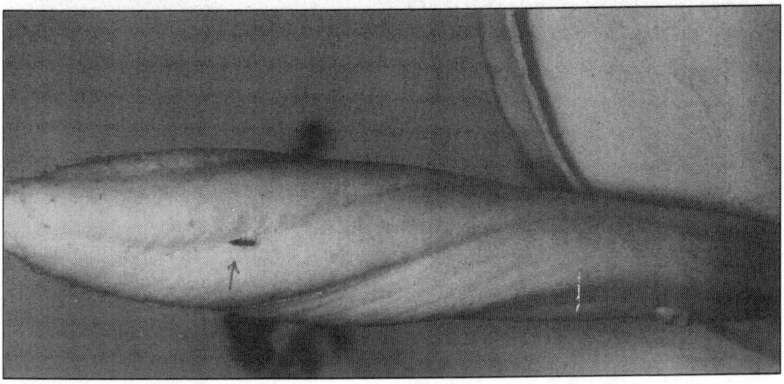

Figure 31: Thrips.

Host Plants

Michelia alba, Michelia champaca, Marigold and other ornamental flowers.

Life Cycle

Eggs are embeded in tender tissues of leaves of Michelia. After hatching the eggs, newly emerged nymphs start feeding on oozing sap of the crop when they rub their mouthparts on tender portions. Nymph moults for 4 times, pupa do not feed at all. Pupal period is 2-4 days. Within 15-20 days life cycle is completed on this crop. Many generations are possible in a year.

Nature of Damage

Nymphs and adults feed on sap of flowers and cause brownish or white spots on flowers. The flowers are discoloured and disfigured affecting market value.

Control Measure

1. Spray the crop with Nuvacron 0.04 per cent or 0.03 per cent Malathion or 0.03 per cent Azadirachtin.
2. Infected flowers be removed and disposed.

Mealy Bug (Figure 32) (*Pseudococcus* sp.)

Mealy bugs are already discussed in previous chapters. However, mealy bugs on michelia cause damage by sucking cell sap and by injecting toxins into the plant body, as a result, leaves become curly, colour becomes dark and leaf surface becomes hard. The pest secrete honey dew like substance on the leaves which create sooty mould on the leaves and affect photosynthesis, growth and yield of flowers adversely.

Figure 32: Mealy Bug.

Control

1. Removal of infested plant parts along with pest stages.
2. Spary the crop with 0.02 per cent phosphamidon or 0.01 dichlorovos.

Scales

White or brown coloured scales suck the cell sap and affect the growth and quality of flowers.

Control

As suggested in Chapter 21.

Snail

Snails belongs to the Class Mollusca. Its body is covered with white shell. They cause damage to Michelia at tender portion by feeding upon that portion.

Control

1. Collection and destruction of snails
2. Spray metaldehyde 5 per cent.

Chapter 21
OTHER PESTS OF ORNAMENTAL PLANTS

Most of insect pests are covered in previous chapters. However, still there are some insects which cause damage to ornamental plants. Their appearance and control is given in this chapter.

Aphids

Appearance

Winged/wingless forms, green, yellow, brown, black, louse litie – suck cell sap and disfigure plants & flowers. Attack many ornamental plants.

Control

Spray rogor 0.03 per cent or malathion 0.03 per cent Malathion is not recommended on fern & petunia as it is phytotoxic.

Jassids (Leafhoppers)

Appearance

Wedge shaped, walk diagonally, yellowish or greenish.

Control

Spary 0.03 per cent Rogar/Malathion (or) carbaryl (sevin) 0.15 per cent.

Scale Insects

Appearance

Circular, semicircular, oval, plate like or bulging bodies sticked to the plant. Black, red, brown, white in colour. Suck cell sap; create sootymould, affect growth & quality of plants and flowers found on several garden plants.

Control

Spray – Nuvacron 0.05 per cent, (or) metacid 0.1–.15 per cent or Temik/Thimet in soil.

Mealy Bugs

Appearance

Small, flat bodied, orange or mahogemy/brownish red with white cottony material on their back. Suck cell sap and affect the growth and quality of flowers of several ornamental plants like *Ephorbia*.

Control

Spray Dichlorovos 0.1 per cent (or) 0.02 per cent phosphamidon.

Leaf Miners

Appearance

Small either Lepidopterous or Dipterous or other insect

Figure 33: Mealy Bug with Flowering Bodies.

larvae. Feed by mining the leaves. Affect growth & quality of flowers.

Control

Spray 0.03 per cent Rogor, 0.05-0.07 per cent Azadirachtin.

Leaf Rollers

Appearance

Lepidopterous larvae (or) sawfly larvae. Fed on ornamental plants by rolling the leaves and thus, affecting growth of the crop and flower quality adversely. Affect many ornamental plants.

Control

Spray 0.03 per cent malathion/0.15 per cent carbaryl.

Ants

Appearance

Social insects shows caste system, division of labour, polymorphism. They cut leaves of plants for food, prepare anthills in lawn, affecting beauty of the garden.

Control

Dusting BHC 10 per cent 20-25 kg/ha

Beetles

Appearance

Various kinds of beetles feed on leaves causing numerous irregular holes by keeping outer edges intact. Feed at night only.

Control

Soil application with Aldrin 0.3 per cent or Chlordane 0.1 per cent.

Weevils

Appearance

Weevils are with snout which is extension part of the head; elongated, with short antenna. Feed on leaves, roots etc.

Control

As above.

White Flies

Appearance

Small moth like, whitish, both nymphs and adults suck the cell sap and affect growth and quality of flowers. Several ornamental plants are affected by this pest.

Control

Spray – 0.05 per cent Nuvacron (or) 0.03 per cent Malathion.

Thrips

Appearance

Slender bodied, small, 1 to 1.5 mm long, narrow winged but fringed. Nymphs are still smaller. Both nymphs and adult rub mouth parts on petals of flowers and other tender parts and feed on oozing sap. Cause white specks on petals, bracts, leaves, etc

Control

Spray 0.04 per cent Nuvacron (or) Apply Aldicarb 550-850 g/100 sqm area (or) 8-10 kg/ha.

White Ants/Termites

Appearance

Social insects, with caste system, division of labour and polymorphism; Digest cellulose by cellulase. Damage various ornamental plants by constructing termateria on crop trunk and feed on woody content.

Control

Spray Aldrin 0.3 per cent (or) Chlordane 0.1 per cent (or) dust 5 per cent above pesticides.

Grass Hoppers

Appearance

Long horned (Figure 34) and short horned grasshoppers. Feed on foliage and affect flowering yield.

Control

Dusting 10 per cent BHC or Sevin 10 to 20 kg/ha. Bird baths attract birds and pick up grass hoppers as food.

Figure 34: Grass Hoppers: Pest of Lawns.

Cutworms

Appearance

Larvae cut seeding of many ornamental plants either in nursery or else were and feed on laid down plants. Thus, cause 100 per cent losses. These are lepidopterous caterpillars with plumpy body and greasy appearance.

Control

Apply thimet at 4-10 kg/ha. Regular ploughing and digging can control pest.

Green Bugs

Appearance

May be elongated or short, suck cell sap from plants and affect growth & flower quality adversely.

Control

Spray 0.03 per cent Rogor.

Bugs

Appearance

Hemipteran bugs snck the cell sap and affect the growth of plant and flower quality

Control

Spray 0.04 per cent Nuvacron (or) 0.03 per cent Rogor.

Caterpillars (Figures 23, 28, 35–36)

Appearance

Caterpillars feed on leaves or flowers of many ornamental plants like marigold, glaricidia, rose, parijatak etc. The caterpillars may be hairy or without hairs. In severe infestation complete skeletonizatron of crop is seen *Spilosoma, Spddoptera, Contarinia maculipennis, Elasmopalpus jasminophagus* spotted worms, semiloopers, scorpion caterpillars, feed on different plants of floricultural importance.

Control

Spray 0.15 per cent carbary (or) 0.03 per cent Malathion.

Cockchafers

Appearance

Beetles feed on foliage at night and defoliate the plant completely.

Control

Apply paradichlorobenzene to soil for repelling grubs from the crop.

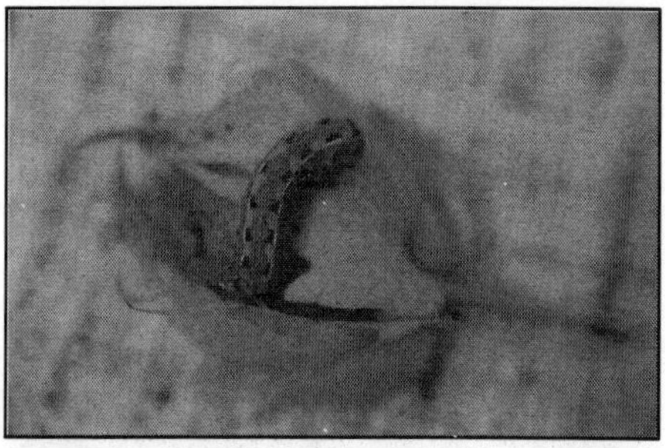

Figure 35: Tobacco Caterpillar.

Figure 36: Leaf Roller Pyroustid Larva.

1. For adults use Nuvacron 0.05 per cent spray.
2. For grubs use Dusting Thimet 4-10 kg/ha

Crikets

Appearance

Crikets feed on leaves of garden plants. They particularly affect lawns by feeding on grasses.

Control

Dust the crop with 10 per cent BHC.

Grubs

Appearance

White grubs are very bad pests of ornamental plants. They feed on roots and affect the growth and quality of flowers and lawns.

Control

Apply thimet 4-10 kg/ha to soil.

Regular ploughing and hoeing is useful.

1. Beetles be collected at night on other plants and destroyed.

Wire Worms

Appearance

Wire worms from Coleoptera feed from underground on roots of ornamental plants and affect the quality and yield of the crop.

Control

As above.

Stem Borers

Appearance

Stem borers are from Lepidoptera and Coleoptera order. The larvae bore into the corm, stem, roots, etc. Beetles also bore. Borers are totally internal feeders and difficult to control with pesticides. They affect growth and quality of flowers. Grass or saw dusts are seen on plants when bored.

Control

Treating stems with Dichlorvos 0.1 per cent (or) Endosulfan 0.05 per cent. Add few drops of kerosene or petrol in the holes and plugging with mud.

Mites

Appearance

Mites are very minute, barely visible to naked eyes. Suck the cell sap and affect the growth and flowers.

Figure 37: Oyster.

Figure 38: Mating *Talicada nicius* (Larvae feed on Brayophyllum).

Control

Spray-Dicofol (kelthame) 0.025 per cent or 0.04 per cent (or) Nuvacron 0.04 per cent or Rogor 0.03 per cent -0.06 per cent.

Nematodes (*Pratulenchus* spp. *Meloidogyne* sp. etc.)

Appearance

Nematodes are whitish, worm like and minute which attack roots of the plants of floriculture. They cause root knots and affect physiology of the plant.

Control

Apply Thimet at 4/10 kg/ha or soil application of Furadan 3 times the quantity of Thimet.

Earthworms

Appearance

Earthworms are many segmented animals which belongs to the class Annelida. These are used in vermiculture but, the worm casts formed by these individuals in fresh lawn look ugly or dirty. Affect the beauty of garden adversely.

Control

Application of chlordane 75 per cent e.c. applied at 30ml/20 sqmt. (or) Granules 5 per cent at 10-20 kg/ha will control worms in lawn.

Slugs and Snails (Figure 37)

Appearance

Slugs and Snails are molluscan pests associated with ornamental plants. They cut the egg plants at ground level and feed on seedlings. Snails acts as scavengers but their presence in the garden is unwelcomed. Slugs and snails attack Ferns, green house plants and flowering annuals. Grey to black coloured slugs are 1.5 to 10 cm. long and are crawlers.

Oyster

Feed on tender parts of crop (Figure 37)

Control

Use poison bait. (250 g wheat bran + 50 g jaggery gur) + 6-8 g metaldehyde) The content is made into mash by adding water. Prepare pellets, and use against slugs and snails.

Garden Tiger Moth (*Arctia caja*)

Appearance

Caterpillars are darkbrown, very hairy, "wooly bears", 70-80 mm long, wander over paths, walks on lawns in June. Moths are with numerous tiger like spots. Larvae found in August-September and hibernates in plant debris, walls, fences, etc. Emerge in early spring and pupate in June. Adults emerge in July/August. Caterpillar hairs irritating. Caterpillars attack garden plants.

Control

Spray 0.15 per cent Carbaryl (or) 0.03 per cent Malathion.

Lily Beetle (*Lilioceris lilii*)

Appearance

Beetles are bright red, 8 mm long and reddish yellow, hump backed larvae covered with black slime. Beetles feed on leaves, stems and seep pods of lilies.

Control

Remove and destroy beetles from April onwards.

Cottony Cusion Scale (*Icerya purchasi*)

Appearance

Adult is 0.25 inch long, marked with ridges by white cottony substance. Both nymphs and adults suck the cell sap.

Control

Spray the crop with 0.03 per cent malathion or carbaryl 0.15 per cent

The Holly Hock Tingidbug (*Urentis euonymus* F.)

Distribution

India, Pak, Bangladesh.

Marks of Identification

Adult bugs are with densely reticulation on the body and wings and lace like appearance hence the name lace bug. Adults are about 5-6 mm long and black in appearance. Nymphs are also blackish, spiny and without wings.

Host Plants

Hollyhock *Althaea rosea* (Linn.), *Sida cardifolia* Linn, *Abutilon indicum* S., *Chrozophora rottleri*, *Ocimum sanctum*, etc.

Life Cycle

Elongated eggs are laid on the leaves from March to June. Eggs hatch within 8 to 10 days. Nymphal period is 10 days to 24 days. There are five nymphal stages. The entire life cycle may be completed on single leaf of the crop.

Nature of Damage

Nymphs and adults suck the cell sap from the under surface of leaves and inject toxins into the plant body as result leaves become curly and pale yellow and then turn brown. Edges turn brown or become dry. Thus, pest affect the growth of crop and quality of flowers.

Control Measures

Spray 0.03 per cent Diamethoate (or)

0.03 per cent Malathion (or)

0.02 per cent Phosphamidon.

The Citrus Psylla (*Diaphorina citri*)

Distribution

India, Japan, Myanmar, China, Sri Lanka, Formosa, New Guinea, etc.

Marks of Identification

Adult is 3 mm, brownish, hing wings shorter than fore wings. Forewings extend well beyond the tip of abdomen. Tail end of body is kept elevated when the insect sits on a surface. Nymphs are flat, orange yellow, louse like and associated with buds.

Host Plants

Plumeria acuminata, Duranta plumieri, Dombeya spectabilis, Citrus, Murraya paniculata, etc.

Life Cycle

Eggs are laid on plants. Incubation period is 4-20 days, nymphal period is 10-20 days. There are five instars in the pest.

Nature of Damage

Both nymphs and adults suck the cell sap from ornamental plants and affect the growth of the plant and look of flowers.

Control

Preventive

Collection and destruction of infested plant parts alongwith pest stages.

Biological

Coccinella septumpunctata, C. rependa, Chilocorus nigrita, Menochilus sexmaculata, Brumus, etc. are lady bird beetles, they feed or nymphs of this pest.

Chemical

Spray the crop with	0.02 per cent phosphamidon
or	0.03 per cent Malathion (or)
	0.03 per cent Rogor.

The Kachnar Psylla

It is distributed in Indian gardens. The nymphs and adults suck the cell sap from *Bauhinia acuminata* Linn. *Bauhinia galphinii* Linn. etc. as a result leaves, flowers and green twigs become rusty, growth is arrested and look of the plant is adversely affected.

Control

1. Preventive – As above
2. Chemical – As above.

AK Butterfly (*Danias chrysippus* L.)

Caterpillars of this butterfly feed on leaves of *Ascleplas curassavica* Linn and skeletonize the plant completely in severe in festation in October and November. They appear on the crop from September to December.

Control

Spray the crop with 0.03 per cent Malathion or 0.02 per cent DDVP.

The Lily Moth (*Plytela gloriosae* Fab.)

Distribution

India, Sri Lanka.

Marks of Identification

Moth contain black hindwings and forewings with mosaic patterns of red, yellow and black. On the apical margin there are black and yellow dots which are arranged in a row. Caterpillars when full grown measures about 38 to 42 mm. They are decorated with black, white and red mosaic patterns. Pupa is obtect type, brown, tapering at one end. Eggs are round and yellow.

Life Cycle

Female moth lay eggs on the apical portion from the under surface of the leaves. Incubation period is 3-6 days. Eggs are laid in clusters of about 15 to 40. Larval period is 15 to 20 days. The larva moults for five times. Full grown larva pupates in soil. Pupal period is 15 to 20 days. Only two generations are possible in a year.

Host Plants

Lily, other lily plants.

Nature of Damage

The caterpillars feed on leaves and skeletonize the plants completely in the garden.

Control

1. Collection and destruction of caterpillars.
2. Ploughing and digging the field for exposing pupae to mortality factors.
3. Spray the crop with 0.15 per cent Carbaryl (or) 0.03 per cent Malathion.

Tobacco Caterpillar (*Spodotera litura* Fab.)

It is also attacks leaves of lily plants. *S. litura* moths have golden zigzag markings on forewings and hind wings are white. Caterpillars are typically velvety yellow with triangular black spots arranged in two rows on the back.

Damage & Control

As above pest.

The Cotton Grey Weevil [*Myllocerus undecimpustulatus maculosus* Des. (Curculionidae – Coleoptera)]

Distribution

India, Pak, Bangladesh.

Marks of Identification

Adults are grey, 3-6 mm long, grubs are whitish and legless, comma shaped and 8 mm long.

Host Plants

Cotton, Chinarose, Rose, Bajra, Maize, Groundnut, Guava, etc.

Life Cycle

Female lays ovoid and light yellow eggs in soil. A single female can lay 360 eggs. Incubation period is 3-5 days.

Larval period is 1-2 months. Pupation takes place in soil. Pupal period is one week. Life cycle is completed within 6-8 weeks and 3-4 generations are completed in a year.

Nature of Damage

Grubs damage roots of ornamental plants and adults feed on leaves of cotton, Chinarose, Hollyhock, etc.

Control

1. Collection and destruction of weevils.
2. Ploughing and digging the field for exposing eggs, larvae and pupae to mortality factors.
3. Spray the crop with metasystox or dimecron.

The Eriophyid Mite [*Aceria jasmini* Chanana. (Eriophyidae: Acarina)]

Distribution

South India.

Marks of Identification

Female – Cylindrical, 150-60 m long. 44 m thick.

Male – 95 m long, rare.

Damage

Mites make webbing on ornamental plants, suck the cell sap from tender leaves, stem, flowers, buds, etc. as a result growth and flower quality and quantity is adversely affected.

Control

1. Collection and destruction of infested plant parts along with pest stages.
2. Spray the crop with Thiodon 0.05 per cent or zolone 0.05 per cent

3. Kelthane 0.04 per cent (or)
4. Nuvacron 0.04 per cent or
5. Rogor 0.03 per cent

Aster Aphid

1. *Macrosiphum artermisiae* – Aphididea – Himiptera]
2. *M. asterifoliae, Aphis fabae* (Black coloured)

Distribution

USA, Europe, India.

Marks of Identification

M. asterifolvae.

Louse like, soft bodied, winged or non winged aphids are cell sap suckers. They are provided with one pair of bars on abdomen and measures about 2 mm in body length.

Host Plants

Asters, and other garden plants.

Infecycle

Reproduction, viviparous and partheno genetic. Many generations are possibly in a year.

Nature of Damage

Both hymphs and adults suck the cell sap from the plants and disfigure the flowers and affect marketability adversely.

Control

1. Collection and destruction of infected plant parts along with pest stages.

2. Spray the crop with 0.03 per cent Malathion (or) 0.02 per cent PHOSPHAMIDON (or) 0.03 per cent Rogor/0.02 per cent Sevin/Metasystox.

Aster Membracid Bug (Figure 39) Membracidae – Hemiptera

Distribution

India

Marks of Identification

Adults are brownish with transparent wings and scutellum is extended in the form of spine like process on the abdomen. Body typically tapers posteriorly. The pronotum also extended as hornlike process on the thorax. Legs are brownish.

Damage and Control

Both mymphs and adults suck the cell sap from Aster plant and affect the growth of Aster and further the quality

Figure 39: Membracid Bug Sucking Cell Sap.

Figure 40: Mealy Bugs.

Figure 41: Aphids.

of flowers. Spray the crop with 0.03 per cent Malathion or Metasystox – R 0.03 Rogor.

Ephorbra Unite Fly (*Dialeurodus dispersu+s* Aleurodidae Hemiptera)

Distribution

India, Sri Lanka.

Marks of Identification

Adults are, small moth like with spiral, filaments with cottony waxy material.

Host Plants

Polyphagous. All mesh and all garden plants in India. Vegetables and fruit crops etc.

Life Cycle

Eggs are deposited on tender leaves. Num,phs suck cell sap and mount for 3 times and become adult.

Nature of Damage

Adults and nymphs suck cell sap and affect quality of flowers by affecting growth of the crop.

Control

As above pest.

Ephorbya Mealy Bugs (*Manococcus citri* and *Pseudococcus* sp.)

(Details already given)

Birds

Many birds visit garden but few of them cause damage to ornamental plants. Important birds damaging garden plants are given below.

Bullfinch (*Pynhula pyrrhula*)

These are small shiny birds with frequent fringes of wood land, coppices and thichkets. Female lays 12-15 eggs in three cluches during May-September. Bird feeds with parties on buds of garden plants. Birds can be controlled by using repellents.

Wood Pigeon (*Columba palumbus*)

Adults are blue grey with a distin of white patch at the base of neck and white bars visible on wings in flight. Nesting peak is during August/September. These birds cause damage by feeding on garden plants from spring to summer. Incubation period of eggs is 3 weeks.

Control

1. Netting and killing.
2. Use repellents.

House Sparrow (*Passer domesticus*) (Passeriformes: Ploceidae)

Female is ash to grayish brown above and fulvus ash-white below. Male is darker above with blackish streaks on wings and black patch on throat and breast and measures about 15 cm long. House sparrow cause damage to leaves and flowers of carnations, chrysanthemum, lawns, lettuce, pears and onion etc.

Control

1. Traping or shooting.
2. Bajra seeds soaked in 2 per cent Fenthion emulsion dried and placed before birds in garden.
3. Spray the crop with Thiurum 0.6 per cent.

The Indian Mynah: *Acridotheres tristis* (Linn.) (Passeriformes: Saturnidae)

Distribution

India, Mauritius, Pakistan, Bangladesh, Myanmar, Sri Lanka etc.

Marks of Identification

Indian mynah is dark brown with bright yellow bill, legs and around eyes. Large white patches present on wings. The bird live in pairs and measures for 22 cm in body length. It lays 4-5 glossy blue eggs.

Nature of Damage

Birds cause damage by feeding upon garden plant seeds and other edible plant parts.

Control

As per house sparrow.

Blue Tit (*Parus* sp.)

This is small bird found feeding on garden plants. It picks camellia, magnolia and other flower buds and flowers for feeding purpose.

Control

As above.

Starling Bird (*Sturnus* sp.)

This small bird feed on lawns. They peck the turf and leave holes and loose tuft of grass. This bird acts as bio-control agent for insect pests in garden and elsewhere.

Control

1. As above.
2. They should not be killed only repelled by various tactics.

The Rose Ringed parakeet (*Psittacula krameri*)

It is commonest grass green bird of India. Attack garden plants. Damage orchid flower, cycas and other flowering and fruiting bodies of ornamental plants.

Control

As above.

The Blue Rock Pigeon (*Columba livia*)

These birds occasionally damage garden plants. They are in parties, settle on lawns, pickup leaves and some time expell excreta.

Control

As above.

Chapter 22
MAMMALIAN PESTS

Rabbit (*Oryctolagus* sp.)

Adults as well as young rabbits feed on lawns, lilies, heathers, hostas, lettuces, etc.

Control

Chemical repellents may be used. They should not be killed and protected by one or another way.

Squirrels (*Funambulus palmarum*)

Distribution

India.

Marks of Identification

It is having 3 median pale stripes on its back. These are noisy creatures with typical hairy tail.

Life Cycle

Male and Female live together for a day or two during

mating period. Gestation period is about 6 weeks. Their young born blind and remain in nest after attaining defendable power they leave the nest.

Nature of Damage

Squirrel constructs a very nice, untidy nest for living and further giving birth to young ones. Therefore, squirrel collect leaves from lawns for construction of nest and can feed on flowering and fruiting bodies in the garden.

Control

Spray 5 per cent zinc phosphide on its bait.

Rats and Mice

There are about 84 species of rodents in India. *Rattus, Mandiota* etc. cause damage to garden by either feeding on garden plants or by cutting them from ground level or by tunneling the lawns. Therefore, rats are bad pests of garden. Rats (omnivorous) *Rattus rattus* is about 16 cm long body unicoloured black to brown dorsally, belly is light coloured, ears are large, tail is longer than body. *Rattus musculus* resembles with black ship rat but are smaller and grey in colour and about 9 cm in body length excluding tail.

Life Cycle

Pregnancy duration is about 3 weeks. A female can produce 8 young ones for about 6-7 times in a year. Young ones start producing within 3 months. A single couple of rat can produce 700 to 800 young ones during a year.

Nature of Damage

Rats feed on garden flowering and fruiting bodies. They make runneling in lawns and affect the beauty of garden.

Control

1. Use traps with poison baits.
2. Use fumigants, attractants, repellents, or chemesterilants for control.
3. Use rodenticides available in the local market.

The Common Wearer Bird [*Ploceus philippinus* (Linn.)]

This bird is found in India, Myanmar, Pakistan and Sri Lanka which resembles with female house sparrow by size and appearance. Dark streaked fulvous appearance on dorsal side and plain unite fulvous from under side are the important features of this bird. The female can lay 2-4 white eggs. It has the breeding period from May to September.

Nature of Damage

The bird cutout leaves of ornamental plants, special lawns are preferred for nesting purpose. They may feed on flowering and fruiting bodies in the garden.

Control

1. Destruction of nests (or)
2. Collection and destruction of eggs.

SPECIES INDEX

S

T

Figure 1: *Macrosiphum euphorbiae.* (Page No. 15)

Figure 2: Damage by Beetles. (Page No. 20)

Figure 3: *Rhipiphorothrips cruentatus.* (Page No. 25)

Figure 4: Damage by Thrips. (Page No. 26)

Figure 5: Damage by Sucking Insects (Thrips). (Page No. 27)

Figure 6: Rose Caterpillars. (Page No. 28)

Figure 7: Rose Cushion Scale. (Page No. 30)

Figure 8: The Cotton Aphid (*Aphis gossypi* Glover). (Page No. 32)

Figure 9: Mealy Bug *Pseudococcus* sp. (Page No. 36)

Figure 10: Mealy Bug *Pseudococcus* sp. (Page No. 36)

Figure 14: Whitefly *Aleurodicus dispersus.* (Page No. 60)

Figure 11: Scale Insects on Stem. (Page No. 39)

Figure 15: White Scales. (Page No. 67)

Figure 16: Mealy Bugs. (Page No. 68)

Figure 17: Marigold . (Page No. 72)

Figure 18: Damage by Thrips. (Page No. 76)

Figure 19: Damage by Thrips. (Page No. 77)

Figure 21: Mealy Bug. (Page No. 82)

Figure 22: Damage by Leaf Miner. (Page No. 82)

Figure 23: Moringa Caterpillar. (Page No. 85)

Figure 24: Larva *H. armigera.* (Page No. 87)

Figure 25: Adult *H. armigera.* (Page No. 87)

Figure 26: Aphids. (Page No. 113)

Figure 27: Bougainvillea. (Page No. 137)

Figure 28: Bihar Hairy Caterpillars. (Page No. 140)

Figure 29: Tuberose. (Page No. 145)

Figure 30: Thrips. (Page No. 160)

Figure 31: Thrips. (Page No. 161)

Figure 32: Mealy Bug. (Page No. 162)

Figure 34: Grass Hoppers: Pest of Lawns. (Page No. 169)

Figure 33: Mealy Bug with Flowering Bodies. (Page No. 166)

Figure 35: Tobacco Caterpillar. (Page No. 171)

Figure 36: Leaf Roller Pyroustid Larva. (Page No. 171)

Figure 37: Oyster. (Page No. 173)

Figure 38: Mating *Talicada nicius* (Larvae feed on Brayophyllum). (Page No. 173)

Figure 39: Membracid Bug Sucking Cell Sap. (Page No. 183)

Figure 40: Mealy Bugs. (Page No. 184)

Figure 41: Aphids. (Page No. 184)